A Second Level Course

Images and Information

Units 9 and 10

The Recording and Reproduction of Visual Information

(An Introduction to Holography)

The Open University Press

THE ST 291 COURSE TEAM

Chairman and General Editor

B. W. Jones

Unit Authors

P. F. Chapman
S. M. Freake
K. A. Hodgkinson
B. W. Jones
I. Lowe
M. Shott
A. J. Walton
A. E. Woolgar

Editors

F. Aprahamian
P. J. Holligan (*Unit 6 and Units 9/10*)

Other Members

K. M. Acharia	(*Photographic*)
G. S. Bellis	(*Technology*)
M. Cawthorne	(*Senior Counsellor*)
D. Coates	(*Monash University*)
A. B. Jolly	(*BBC*)
P. R. Lefrere	(*IET*)
J. K. McCloughry	(*SCS*)
L. R. A. Melton	(*Library*)
A. J. Millington	(*BBC*)
P. Noskeau	(*Course Assistant*)
E. F. Smith	(*BBC*)
S. J. Swithenby	(*Physics*)
J. Wilkinson	(*Capricornia Institute of Advanced Education*)

The Open University Press
Walton Hall, Milton Keynes
MK7 6AA

First published 1977
Second edition 1978, reprinted 1981

Copyright © 1978   The Open University

Designed by the Media Development Group of the Open University.

Printed in England by
Staples Printers St Albans Limited
at the Priory Press.

ISBN 0 335 04380 1

This text forms part of an Open University course. The complete list of items in the course appears at the end of this text.

For general availability of supporting material referred to in this text, please write to: Open University Educational Enterprises Limited, 12 Cofferidge Close, Stony Stratford, Milton Keynes, MK11 1BY, Great Britain.

Further information on Open University courses may be obtained from the Admissions Office, The Open University, P.O. Box 48, Walton Hall, Milton Keynes, MK7 6AB.

2.2

Units 9 and 10　　**The Recording and Reproduction
of Visual Information
(An Introduction to Holography)**

CONTENTS

Prelude

This two-Unit Block is subtitled 'An Introduction to Holography'. But what is holography? What does a hologram do? If you've ever viewed a reconstructed holographic image, then you're probably already curious to know how holograms work. But if you haven't, then I'd like you—before you embark on the explanatory details in the Main Text—to indulge in a bit of pre-study experimentation. I'd like to invite you to view the 'image' stored on the hologram provided in the Home Kit.

The hardware you require is minimal: the optical bench; three saddles and a pillar; the laser; the diverging lens A; and of course, the hologram (which is labelled 'Time and Space'). To reconstruct the holographic image, you should proceed as follows.

Mount the laser at the far end of the optical bench, with the beam directed like a horizontal plumb-line straight along the rail. Mount the hologram in the optical pillar, and position it in a saddle at the other end of the bench. Adjust the height and transverse position of the hologram so that the laser beam strikes it approximately in the centre. The words 'Time and Space' should be at the top of the hologram, facing *away* from the laser. Now place the 'magnetic' lens A on the front of the laser box so that it expands the beam into a cone of light. Adjust the position of the lens until the laser light covers the whole area of the hologram. Now, standing at the hologram-end of the optical bench and looking towards the laser, twist the right-hand side of the hologram away from you (i.e. an anticlockwise rotation as seen from above) by about 30 or 40 degrees. If necessary, push the saddle holding the hologram a little nearer the laser, so that the laser light just fills the hologram. Now look *through* the hologram (as though it were a window) from the side facing away from the laser. The photograph in Figure 1 should give you the idea.

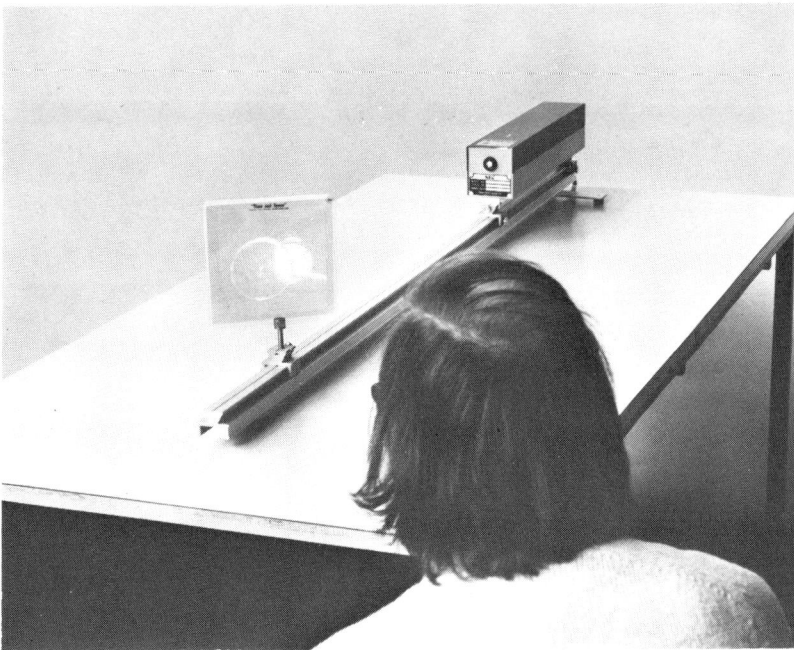

Figure 1 Viewing your Home Kit hologram.

Move your head about and look through the hologram from as many different angles as possible—from the left, from the right, from above and from below. (The laser light is quite harmless when diverged by the expanding lens A.) Is there a position that enables you to see all the figures on the stopwatch magnified? Is there a position that enables you to see all of the stopwatch without looking through the magnifying glass?

Or try this. Put your hand behind the hologram in a plane coincident with the magnifying glass, and then in a plane coincident with the stopwatch. How far apart are these planes?

So now you know what a hologram does. Are you impressed?

Table A

List of scientific terms, concepts and principles used in Units 9 and 10

Taken as prerequisites		Terms developed in this Unit	Page No.
Introduced in a previous Unit	Unit No.		
amplitude	WR*	Bragg 'reflection' (diffraction)	64
amplitude transmittance	3/4	camera obscura	11
aperture stop	1,6	carrier fringes	41
bandwidth	5	character recognition	69
characteristic curve	2	coherent superposition	26
coherence area	5	colour holography	61
coherence length	5	computer holograms	69
coherent illumination	5	cross-talk	62
convex lens	WR, 6	'dummy-head' visual recording	14
depth of field/focus	1	Fourier transform hologram	25
depth of modulation	1	Fresnel hologram	41
diffraction	3/4	Fresnel transform	22
distortion	2	fringe resolution	47
exposure time	1	hologram bleaching	61
F-number	1	holographic data-storage	69
film speed	2	holographic interferometry	67
Fourier transform	3/4	holographic life-studies	48
Fourier transform plane	3/4, 6	holographic movies	69
Fraunhofer diffraction	3/4	holographic TV	70
Fresnel diffraction	3/4	holography	20
interference	WR	obscuration (overlapping)	10
linearity (of film)	2	optical field	20
magnification	WR, 6	parallax	18
Michelson interferometer	5, TV5	perspective	10
modulation	1	phase hologram	61
optical diffractometer	6, TV6	pseudoscopy	39
phase	WR, 3/4	reflection hologram	67
phase-contrast	7/8	side-band Fresnel hologram	41
phase grating	7/8	spectral sensitivity	48
phase object	7/8	stability requirements	48
pulsed lasers	Radio 3	stereophotography	14
real image	WR	volume hologram	62
resolution	6		
retina	1		
sinc function	3/4		
spatial coherence	5		
spatial filter	7/8		
spatial frequency	1, 3/4		
spectral sensitivity (of film)	2		
standing wave	WR		
temporal coherence	5		
temporal frequency	WR		
time-averaged intensity	2		
virtual image	WR		
wavelength	WR		
X-ray diffraction	3/4, 5 and TV6		

Waves and Rays.

Objectives

After studying these two Units, you should be able to:

1 List and recognize those depth cues which can be incorporated into a conventional *two*-dimensional representation of a three-dimensional object or scene. (SAQs 1, 2 and 3)

2 Enumerate those depth cues which *cannot* be so incorporated, and provide examples of modifications to the photographic technique which attempt to compensate for their absence. (SAQ 4)

3 Explain why it is that the holographic process is capable of storing *all* the information relating to object depth.

4 Explain why a hologram can be made in any plane in an imaging system.

5 Describe how, and explain why, the addition of a coherent background preserves the phase component in an optical field. (SAQs 6 and 7)

6 Calculate the typical carrier-fringe spatial frequency given the average angle between the reference beam and object beam in a holographic set-up (and vice versa). (SAQ 8)

7 Calculate the typical depth of modulation of the fringes at any part of the hologram, given the ratio of the reference and object light intensities at that position. (SAQ 10)

8 Sketch and explain the relationship between the object position, and the positions of the reconstructed real and virtual holographic images, relative to the plane of the hologram. (SAQs 9, 11 and 14)

9 Describe the effect on the two (real and virtual) holographic images of playing back only a small portion of the hologram, and explain how the effect differs for different types of hologram (a Fourier transform hologram compared with a Fresnel hologram for instance). (SAQs 13, 20, Radio 5 and TV9)

10* Explain in what way the resolution capabilities of the holographic film limit the quantity of information about the object which can be stored in (i) a side-band Fresnel hologram, and (ii) a Fourier transform hologram. (Radio 5)

11 Describe the experimental conditions necessary to make satisfactory holograms. (The holography experiment in Section 6, TV9)

12 Make your own holograms. (The holography experiment in Section 6).

13 Describe the problems involved in making a full-colour hologram (SAQs 15 and 16) and explain how these problems can be overcome (SAQ 17).

14 Describe the problems involved in playing back holograms with white light, and list some possible solutions. (SAQs 18, 19 and TV9)

15* List and describe the various techniques of holographic interferometry, and explain how these techniques can be used to determine movement, deformation and stress, or optical density changes of the object being studied. (TV 10)

16 Summarize the other advances in holography discussed in Section 9 of these Units.

17* Explain the origin of laser speckle. (TV 10)

*Certain Objectives have in brackets after them a TV or radio programme number. This means that the TV or radio programme supports that part of the Unit to which the Objective refers. The Objectives marked by an asterisk, however, refer to assessable material covered only in the broadcast(s) indicated. You will *not* find this material in the Main Text.

Introduction

In these two Units I want to try to give you some understanding of the methods of recording and replaying visual information. The ultimate goal is, of course, a description of the mechanism of holography. But rather than limit the discussion to holography alone I want to take you on a tour of other visual-recording techniques. In this way I hope to develop not only the mechanism of, but also the rationale for, holography. So *en route* we shall consider painting and drawing, photography, stereophotography and even movie-films. We shall try to understand the relevance of perspective, binocular vision, eye-focus and head-movement to three-dimensional perception. By the time you reach the technicalities of holography itself, you should have a very clear idea of what has to be achieved in order to make good the deficiencies of the other recording methods.

But a theoretical understanding is not everything. Holography is a very practical art, and one in which we should like you to gain some expertise. So we have designed a special 'holography box' which, when used together with the high-resolution holographic film (provided as two ten-exposure cassettes), should enable you to make your own 'three-dimensional' holograms without too much trouble. However, do try to complete the whole of the 'making your holograms' part of the experimental work in *one* session. You are likely to waste time, and produce poor results by tackling it in bits and pieces. Due to the fairly high experimental content of these two Units, you may find distributing your time sensibly a little more difficult than usual. I suggest that you should try to reach the end of Section 4 within your first ten hours of study. (Incidentally, Section 4 is probably the most difficult Section in these two Units, so don't despair if you find the 'going' a bit tough around there!) Your second 'Unit's worth' of time should be spent on the remaining Sections, allowing about six hours altogether for *all* the experimental work.

1.0   Before holography — *Conventional image recording techniques*

1.1   Art — a perspective view

How do we make our visual experiences permanent? How can we best record for others, who might be separated from us in either space or time, those images which we ourselves have seen, and which we would also like them to see? These are questions which have exercised the minds of men from the very earliest times (think of the wall-drawings produced by the cavemen for example) right up until the present day. But over the centuries the techniques and principles have changed, and what I would like to do in this Section is to explore how and why these changes have taken place.

Of course, if we are not careful, this kind of approach can quickly lead us into philosophical arguments questioning the nature of 'artistic perception'. So let me say at the outset, that I am only concerned here with the techniques of *visual fidelity*—with the methods which allow us to regenerate an image which bears the closest possible resemblance to the original 'visual object'. This is not meant to be a denigration of artistic licence, but rather an attempt to separate the mechanics from the artistry. For, if we understand the tools at our disposal, we are obviously in a better position to use them with taste and imagination. And I, for one, am convinced that the artistic and imaginative use of holography is just as possible (and desirable) as the creative use of paints and canvas!

Yet it may come as a surprise to you to realize that even the first steps along the road to realism in art were taken only comparatively recently. Look at Figure 2. This drawing is by an early fifteenth-century artist. Would you say that the picture was an accurate representation of the scene as it might have appeared to an onlooker? Well obviously not, but can you say why?

Figure 2 'Glass Making'—one of twenty-eight miniatures illustrating the travels of Sir John Mandeville (fifteenth century). Why does the scene depicted in this drawing look artificial?

This might sound a very trivial question to ask nowadays; after all, it's quite clear that the artist had his *perspective* all wrong. Yet the discovery of perspective didn't pass into the art world until the fifteenth-century, and then it engendered such excitement (particularly in northern Italy) that the adherents to the new technique were able to form themselves into the new school of the *Perspectivi*.

In retrospect, we can see that the physics of perspective is very simple. The size of an image formed on the retina of the eye depends on the (solid) angle that the object subtends at the eyelens. Furthermore, for a specific object, the solid angle decreases as the object is moved further away (Fig. 3).

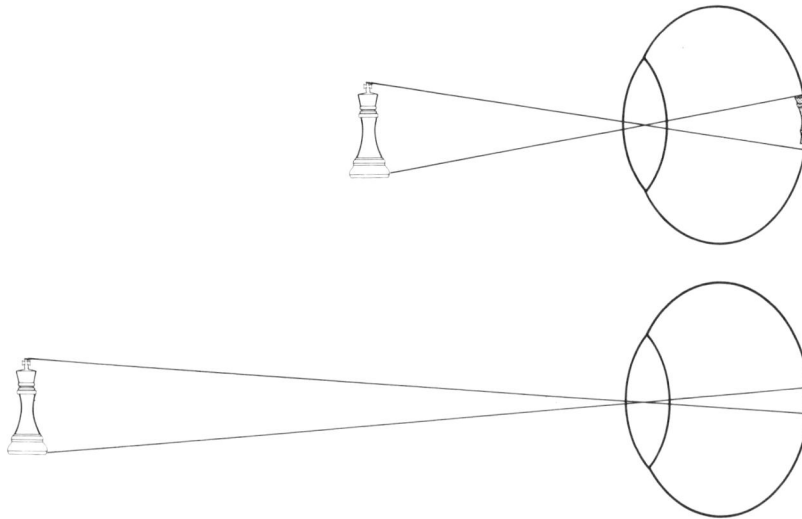

Figure 3 The more distant an object is from the eye, the smaller is the image on the retina.

SAQ 1 The chesspiece in Figure 3 is 5 cm high. Calculate by what factor the size of the image on the retina is reduced when the chesspiece is moved from a position 25 cm away from the eyelens, to a position ten metres distant from the eyelens.

Image size, or perhaps more commonly, *relative* image size, is therefore an important perceptual cue to object distance. But a painting or drawing is necessarily a mapping of a three-dimensional world onto a two-dimensional surface. So, in order to convey the impression of distance, the artist must make use of optical illusions. He must make distant objects smaller than near ones; he must make receding objects become smaller as they go away from us. (Fig. 4—which is a painting, not a photograph—shows both these effects very clearly.*) The artist must also distort the true shape of objects which have depth, so as to simulate, on paper, the distorted image of these objects that the eyelens forms on the retina. As Figure 5 demonstrates, some artists took up this particular challenge with a vengeance!

So these were the 'tricks' which had to be learnt. Any artist worthy of his name at this time, had to be master of this perspective illusion. If you doubt that such tricks really do generate an illusion, then Figure 6 should dispel those doubts. Here, Escher, with twentieth-century hindsight, has used the illusion and turned it back on itself. If you look at a localized area of the drawing, your brain has no difficulty in recognizing the convention of the two-dimensional mapping, and translating it back into a three-dimensional 'thought-image'. But because the perspective codings in different regions of the drawing are not consistent with each other, your brain is not able to resolve the content of the whole scene. Instead, you are left with the disquieting thought that perhaps water can run uphill after all!

Of course, perspective is not the only depth illusion that the artist can make use of. Perhaps the most obvious depth cue is *obscuration* or *overlapping*. If one object obscures, or overlaps another, then the obscuring object must obviously be in front of the obscured object. The chandelier on the right of the picture in Figure 4, for instance, is clearly in front of the pillar it partially hides.

At the other extreme, perhaps the most subtle depth cue in the artist's repertoire is the one afforded him by the careful use of light and shadow. I feel, for instance, that de la Tour, in his painting of 'St. Joseph the Carpenter' (Fig. 7) manages to generate a quite uncanny sense of depth by the masterly way in which he uses the light from the candle to throw the foreground into relief.

*Figures 4, 5 and 7 can be found on the colour plate between pages 12 and 13.

Figure 6   'Waterfall' (1961) by M. C. Escher. Spot the deliberate mistake!

SAQ 2   The early fifteenth-century artist of the drawing shown in Figure 2 almost certainly never had as one of his aims, a desire to represent the scene as the eye would see it. (A god's eyeview was often considered more appropriate.) Yet, if this had been one of his goals, he would certainly have had to depict several things in the scene rather differently. Identify as many of the things that would require modification as you can, and say in what way they would have to be modified.

1.2   Photography

Unfortunately, not all artists found the technique of perspective drawing easy to acquire. Considerable skill was required for instance, to scale down the size of an object and still keep the angular perspective correct. So it is not surprising to find at this time, a growth of interest in mechanical and optical aids designed to help the artist capture a 'true perspective likeness'. Dürer, for example, describes how he placed his model behind a frame supporting a network of strings, the strings being so arranged as to divide up his view of the model into an array of equal squares. He then, with the aid of a sighting tool, proceeded to draw the outline of his subject onto squared paper.

It was at this time also that the *camera obscura* (literally a 'dark room') became popular as an aid to perspective drawing. Figure 8 shows how the rays of light from the illuminated object are admitted to the room through a small hole in one wall. These rays then strike the facing wall in such a way as to form a faint, inverted image of the object upon it. The artist would sit in the room and copy the (now two-dimensional) image onto his canvas*. In some versions of the camera obscura, the

camera obscura

*Perhaps some of you will recognize the camera obscura as nothing more than a form of large pinhole camera.

rays would be intercepted by an angled mirror, positioned so as to re-invert the image and reflect it onto a horizontal working surface. The artist could then simply trace around the image.

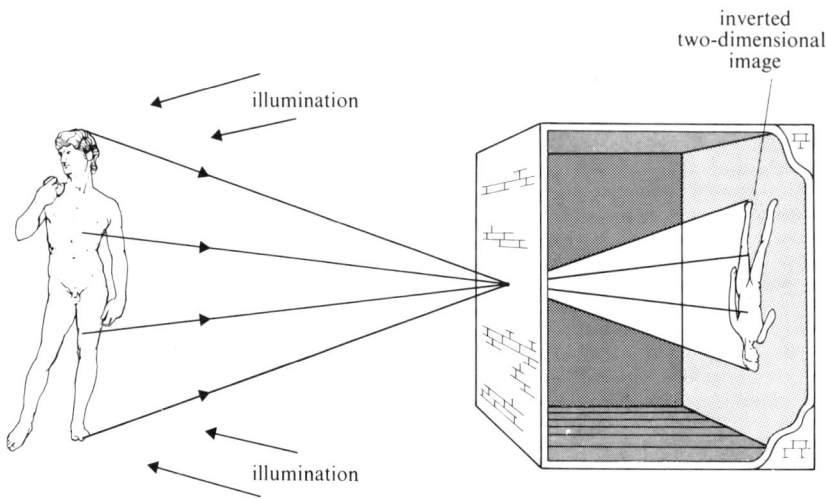

Figure 8   The camera obscura was used as an aid to perspective drawing. If the hole which admits the light to the chamber is small enough, a single point on the object will (approximately) correspond to a single point on the image. This image is faint because the amount of light admitted is proportional to the area of the hole. Increasing the size of the hole will blur the image.

Not surprisingly, it wasn't long before a portable, more useful, version of the camera obscura was developed. Here, the image was displayed on a translucent screen, so that the artist could see his picture from outside the box. The trouble with this system was that the faint image was now even more difficult to see. As Figure 9 shows, however, when a convex lens is used in place of the pinhole, the image brightness is greatly increased without any loss of image sharpness.

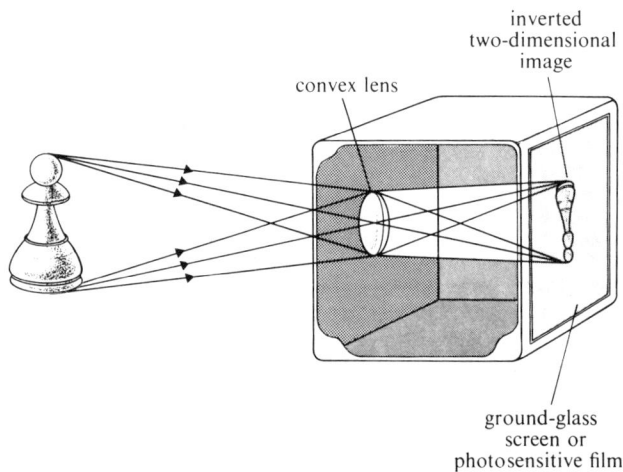

Figure 9   A portable version of the camera obscura makes use of a ground-glass screen and a convex lens. Since the amount of light admitted is proportional to the area of this lens, the image brightness is greatly increased without any loss of image clarity. A single point on the object still gives rise to a single point on the image (provided that $1/v + 1/u = 1/f$) because of the focusing properties of the lens.

The device shown in Figure 9 bears a remarkable resemblance to our modern camera. Yet as it stands it can still only be used as an aid to drawing—the artist still has to copy the image. But with the development of photosensitive materials (almost exactly 150 years ago), even this step was taken out of the artist's hands. The image could now be permanently captured directly on the photosensitive plate. The recording of images had become an automated science. Correct perspective could be achieved by anyone with access to a film-camera.

12

Figure 4 'Interior of a church' by Emmanuel de Witte (ca 1615–92) Notice how the artist has mastered the technique of perspective—the columns and arches get smaller and narrower as they recede into the background.

Figure 7 'St. Joseph the Carpenter' by Georges de la Tour (1593–1652). Light and shadow are clearly important depth cues.

Figure 5 'Cristo morto' by Andrea Mantegna (c. 1431–1506). An extreme example of distortion of shape for realistic effect. It is hard to appreciate that the body is drawn approximately as tall as it is wide.

There has been an enormous improvement in camera and film technologies since those early days. We now take for granted the fact that the camera will give us a perfect recording of any scene, complete with all the depth cues that an artist could have incorporated in his paintings. The photograph in Figure 10, for example, contains *all* the depth cues I have so far mentioned.

Figure 10   The film camera provides automatic perspective.

SAQ 3   Give examples of the various depth cues that are apparent in this photograph (Fig. 10).

But this is by no means the end of the story. Photographs might produce a good *likeness*, but in no way can you be fooled into thinking that a photograph of an object is the object itself. Perspective-related depth cues alone are obviously not sufficient to generate a totally convincing impression of three-dimensionality. So let us look at this problem in more detail. Let's see if we can draw up a list of factors which might allow us to differentiate a real-world object from its photographic image.

1.3   Binocular vision

Since the camera intervenes in the system when we take a photograph, it might be prudent to ask ourselves whether the camera adds or removes anything of importance from the image. Well, the camera lens is not unlike our eyelens; the film as a detector is quite a good substitute for the retina, particularly if we use colour film; and the aperture stop seems to make a pretty good imitation eye-pupil. In fact, everything considered, the camera is not at all a bad attempt at a mechanical eye. But *we* have not one, but *two* eyes. Does this matter? Well, it turns out that it does. Since our two eyes are separated in space (usually by about 7 cm), the viewpoint of each eye is obviously slightly different. This means that there will be a slight difference between the two retinal images—a difference which can be analysed by the brain when it compares these images. Furthermore, if an object is a long way off, then the angular difference between the two viewing directions is small, and consequently the difference between the two retinal images is also small. For a close object, however, the angle between the two viewing directions is much larger (Fig. 11) and the corresponding difference between the two retinal images is now very much more noticeable. So the brain can make use of this disparity between the two images to infer the relative distance of objects, and hence the three-dimensionality of a scene.

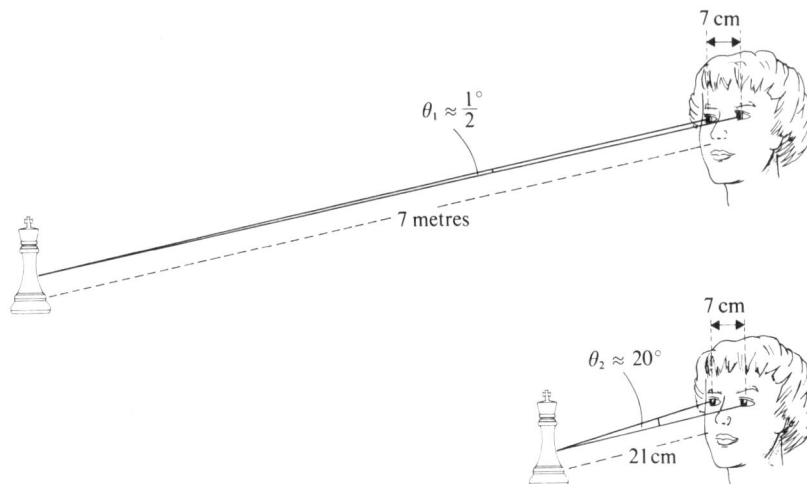

$\theta_1 \approx \frac{1}{2}^\circ$

7 cm

7 metres

$\theta_2 \approx 20^\circ$

7 cm

21 cm

Figure 11 The closer an object, the greater is the disparity between left and right retinal images.

1.4 Dummy-head recording

So, since binocular vision seems to be important for depth perception, let's try to adopt a rationale for visual recording which takes this into account. We'll label the technique we are going to use *'dummy-head' visual recording*. The philosophy is as follows.

'dummy-head' visual recording

If we can't be present at the actual event as it happens, we will substitute a visual recording machine which duplicates, as nearly as possible, our human visual detection system. Let us build a stand, and mount two cameras on it with their lenses separated by about 7 cm. We now have a kind of dummy-head as far as visual recording is concerned. After all, as we pointed out earlier, the camera lenses are not too different from our eyelenses—and photographic film operates in a way which is quite similar to the action of the retina (both processes depend on a photochemical reaction in the first instance to activate the detecting device). Of course, the eye then goes on to convert the chemical reaction pattern into electrical data, which is then sent via the optic nerve to the brain. The developed film, on the other hand, retains the information in a pattern which bears a more or less one-to-one correspondence with the distribution of information on the retina. But with our dummy-head we get not one photograph, but two; and providing that we have been sensible with the positioning of our 'head', these two photographs will correspond to the two retinal images which our eyes would have produced had *we* been present instead of the two cameras.

If we now devise a playback system for the two photographs such that only the left retinal-like image goes into the left eye, and only the right retinal-like image goes into the right eye, then we might expect to perceive the extra depth information provided by the binocular viewing. We might expect the scene to look 'more' three-dimensional.

Actually this sort of photography has been around for a long time now—it's known as *stereophotography*. In practice you don't need two cameras, one will do. You simply take one photograph from a left-hand viewpoint, and then take a second photograph from a right-hand viewpoint. In fact, you can buy a special adaptor which, when attached to the top of a tripod stand, allows you to mount the camera in such a way that it can be moved accurately from the left- to the right-hand position.* If the two developed slides are then viewed through a double-barrelled slide viewer—hey presto!—you see a three-dimensional image. But don't just take my word for it; have a look at the 'stereopair' of photographs in Figure 12, opposite. I have explained in Figures 12 and 13 how you can view the chessboard

stereophotography

*If you are living on a tight budget, you might be interested to know that you can do almost as well without any mechanical aids whatsoever! With a little bit of practice, you can soon acquire the knack of putting just the right amount of weight on your left foot when you take the first photograph, and then just the right amount of weight on your right foot when you take the second photograph, to give you a viewpoint-spacing of about 7 cm.

Figure 12   A stereoscopic-pair of photographs. Hold a plane mirror along the right-hand side of your nose and vertically above the dot between the photographs (see Fig. 13). Look at the left-hand photograph directly with your left eye, but with your right eye, view the right-hand photograph by reflection from the mirror. Adjust the angle of the mirror until the left and right images coincide. You will probably find it easier to concentrate on only one of the chesspieces (one near the centre of the board) to align the images, and then to allow your vision to expand to encompass the whole of the board. (Incidentally, it's white to move, and mate in six!)

stereoscopically by using the plane mirror in your Home Kit to direct the right-hand photographic image into your right eye, while looking at the left-hand photograph directly with your left eye. (Incidentally, we had to reverse the right-hand photograph of the pair at the printing stage, so as to compensate for the lateral reversal which takes place when the photograph is viewed through the mirror.)

It is quite interesting to see just how much the two photographs in this stereopair differ from each other. (It gives us some idea of how different the two retinal images would have been.) In Figure 14, the two photographs have been superimposed; the different viewpoint of each camera is very obvious.

Figure 13   How to view the stereopair of photographs in Figure 12. You can use the plane mirror in the Home Kit.

Figure 14   The disparity between the two stereophotographs of Figure 12 is very obvious when the two photographs are superimposed. The common focus of both left and right cameras was the bishop in the centre of the 'back row'.

2.0   Holography: A Philosophy — *What's wrong with photography?*

2.1   Depth of field

I don't know how convincing you found the three-dimensional effect of the stereophotographs in Figure 12. I personally feel that although the result of stereophotography is often very pleasing, there is still something very stilted about the scene which detracts from the sense of realism.

Of course, we might expect there to be some lack of fidelity because of the way in which the photographs are made and replayed. For instance, having simulated our eyes by using a 'dummy-head' recording technique, it's unfortunate, to say the least, that we have to feed the information back through our eyes again on replay. Ideally we would have preferred to convert the information on the two photographs into electrical impulses, and then feed these impulses directly into the optic nerve—but that's a bit tricky at present!

But surely that's not all that's wrong with stereophotography? Perhaps there are other factors in addition to binocular vision, which serve to differentiate the photograph from the live situation?

Look at the two photographs in Figure 15. The object depicted is the same in both cases—but the photographs don't look the same. Why not? Well obviously a lot of the second photograph is 'out of focus', and generally speaking you might reasonably prefer the other photograph in which everything in the picture is more or less in focus. But is this true of life? Let's try an experiment. Read the following paragraph of text through first, and then try the simple experiment described there for yourself.

16

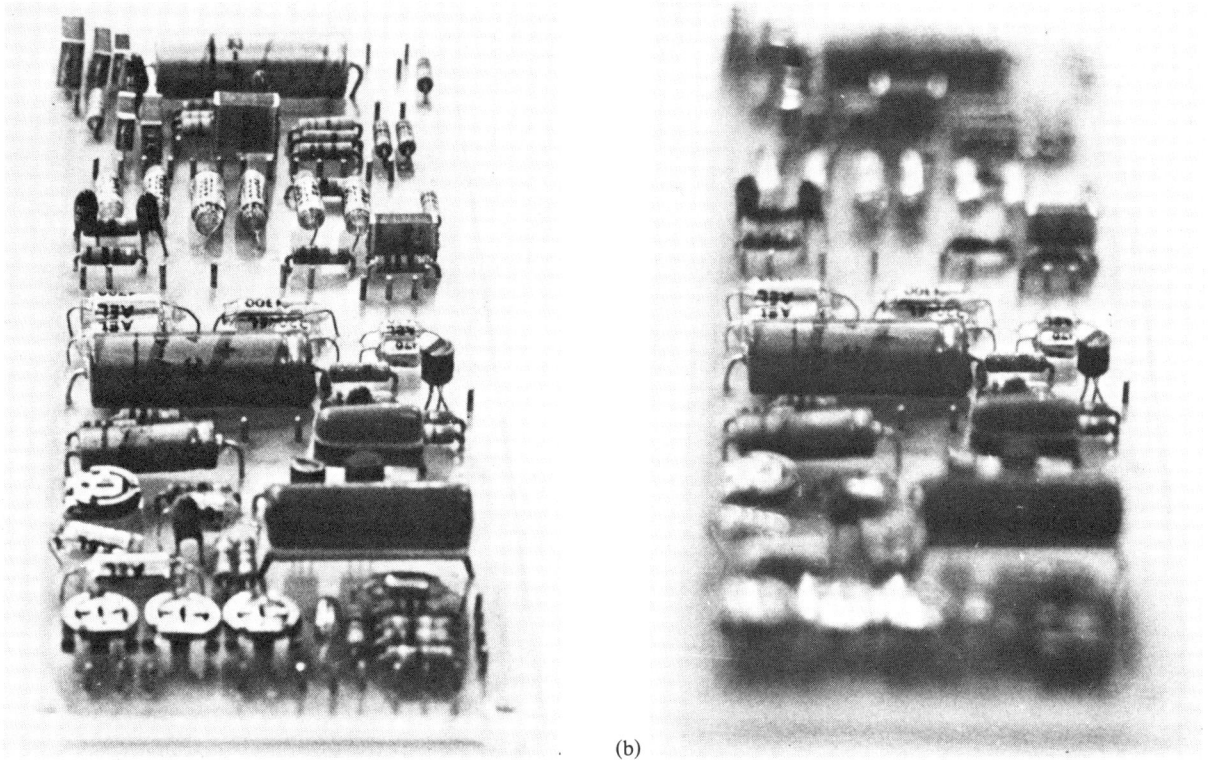

Figure 15 These two photographs, of electronic components on a printed circuit board, were taken under identical lighting conditions, and with the camera focused on the same point (somewhere near the middle of the board) in both cases. But photograph (a) has a large depth of field (it was taken with an aperture size of $F/8$ and an exposure time of $\frac{1}{4}$ second), whereas photograph (b) has a small depth of field (taken at $F/1.4$ and $\frac{1}{125}$ second).

Prop up this text on your table, or desk, at a distance of about 1 metre (3 feet). Take a sharpened pencil (or a ballpoint pen), and hold it, point up, in front of one eye. Hold it as close as you can to your eye, but not so close that you can't focus clearly on it. You will probably find that a distance of about 15 cm (6 inches) is about right. Close your other eye. Now line up the text, the pencil-point and your eye, as shown in Figure 16. Focus first on the pencil-point. Without changing your focus, allow

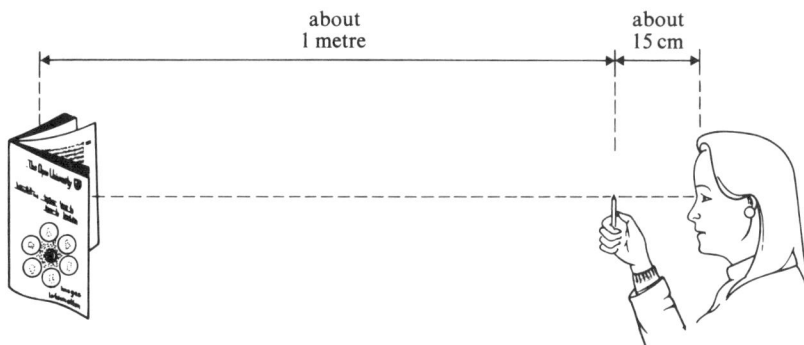

Figure 16 Position the pencil-point in line with one of your eyes and this text placed about 1 metre away from you. Can you focus on both the pencil-point and the text at one and the same time?

yourself to become aware of the text in the distance. Can you read the text? Now try it the other way round. Focus on the text, and then allow yourself to become aware of the pencil-point in the foreground. Make sure that the text stays in focus. Can you see the pencil-point clearly? *Try this experiment now, and then read on.*

The important conclusion to draw from this experiment is that objects at different distances are not usually in focus simultaneously. In fact when we switch our attention from a near to a distant object we also change the focal length of our eyelens. Because we do this subconsciously it's easy for us to forget about it, or not even realize it's happening at all.

But whether the process is subconscious or not, whenever we look at an object, the image signals coming along the optic nerve can be correlated by the brain with the signals generated by the eyelens muscles in such a way that the very act of focusing on the object can be a clue to the object's distance.

So, although most of you might have preferred the first of the two photographs in Figure 15, and although this is the kind of photograph photographers most frequently need to take, the second photo is a much better example of the way in which the eye works. Since we were extolling the virtues of the camera as a mechanical eye not so long ago, we ought just to remind ourselves how we managed to take both kinds of photograph in Figure 15.

In each case the camera lens was 'focused' on the same point in the middle of the circuit board. The photographs only look dissimilar because of the different aperture sizes, and the different exposure times used. Figure 15(a) employed a small aperture and relatively long exposure time, whereas Figure 15(b) was taken with a large aperture and a correspondingly shorter exposure time. (The total amount of energy falling on the film was then about the same in both cases.) You probably know that the larger the aperture size, the smaller is the depth of field i.e. the smaller the range of distances that will appear in focus on the photograph.*

You might ask at this stage why the eye can't achieve a large depth of field simply by stopping-down the pupil size. The answer is that it can, but unfortunately not on command. Pupil size is directly determined by light brightness (as a protective measure) and we simply have to accept the depth of field which corresponds to the brightness of the particular scene.

Hence, we must now add 'lack of focusing cues' to our list of what's wrong with photography. And what's more, this problem is not cured by stereophotography either. For even here, when we view the stereopair, we focus on a single plane only, namely the plane of the two photographs. If the photos were taken with a large depth of field, then everything is in focus without our having to change the focal length of our eyelenses. If, on the other hand, only a small depth of field appears on the photos, so that some parts of the scene are out of focus, no amount of refocusing our eyes can bring those regions back into focus.† In this case the only focus position we are allowed was decided for us at the time the photograph was taken. A very unsatisfactory state of affairs.

2.2 'I can't see — please may I stand up?'

Perhaps the biggest drawback with photography is the fact that the photographer has to pick one viewpoint per photograph. Once the photograph has been taken the viewpoint is fixed. It is no good realizing when you look at the photograph that John was standing right in front of Mary, obscuring her from view. In the live situation of course, you could have moved your head slightly, and so been able to see both John *and* Mary. But with the photograph, Mary's fate was sealed the instant the shutter clicked!

An alternative approach would be to try to build some kind of movement into the photograph. For instance, we could use a movie camera. But this is cheating a bit, because we now have not one, but several photographs. Yet even this technique still has its drawbacks. For instance, since you weren't necessarily making the movie, you're not sure in which direction the camera was moved, how far it was moved, or how fast. As the viewer, you can only guess at these factors from the way in which the scene changes. In real life however, the retinal images can be correlated with the motor-mechanism which you use to move your head—a much better way to go about things.

There is another important point to be made here as well. Head movement, by allowing you to change your viewing position, allows you to 'look round' obstacles (as we've just seen). But head movement is also another important cue to distance perception, through the mechanism of *parallax*. This is the term used to describe the illusion that, when two objects are more-or-less in line with one of your eyes, the more distant object appears to move in the same direction as that moved by your head (and hence your eye), whereas the nearer object appears to move in a contrary

parallax

*We mentioned this in the Home Kit Handbook, and again in Unit 1.

†Unless we indulge in the kind of 'image-deblurring' procedures described in Units 7 and 8.

direction (see Fig. 17). Hence, without head movement we cannot have parallax, and without parallax we lose yet another visual cue to distance and three-dimensionality.

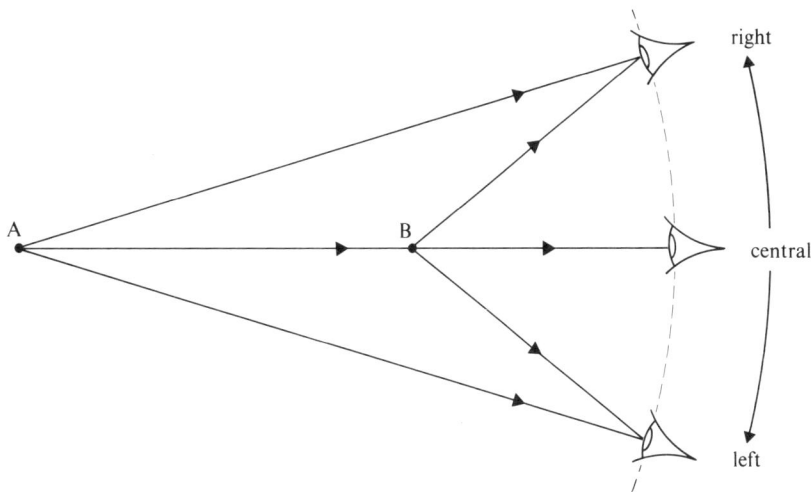

Figure 17 You do not need two eyes to see parallax effects—they are apparent with one eye closed. When you position your eye along the line AB, then object A is located directly behind object B. When you move your head to the right, however, the line of sight to A is further to the right (i.e. in a more clockwise direction) than the line of sight to B. Alternatively, when you move your head to the left then A appears to the *left* of B. A comparison between the directions to A and B seems to suggest that A moves in the *same* direction as your head, whereas B moves in the *opposite* direction.

2.3 Stereophotography does not solve all the problems

It would seem therefore, that although (static) stereophotography can reproduce binocular vision, it nevertheless still has at least two other failings—absence of 'head-movement' and lack of variable-focus distances. Both these parameters are fixed at the time of recording, and on playback we have to accept this limitation whether we like it or not. There is no easy way to go back and ask what exactly the writing on that notice in the distance says, when the camera was focused on the foreground at the time the photograph was taken. But this is exactly the kind of eye-focus readjustment which we make, and need to make, all the time in a live-viewing situation. Even when you look at a movie-film, you have to be satisfied with the head-movement responses which the cameraman bequeathed to you when he made the film. And what's the betting that as far as you're concerned he pointed his wretched camera in the wrong direction! If we are going to compensate for these failings it rather looks as though a different approach is called for.

> **SAQ 4** List the depth cues *not* present on a conventional photograph. What other photographic techniques can be used to try to overcome the absence of these cues? How well do they succeed?

2.4 Amplitude and phase — a first stab

At this stage I would like to digress for a short while, in the hope that by so doing I can better point you in the direction of a solution to our problems.

What I want to do is reflect on the importance of phase in the visual perception process. We know that light waves can be characterized at any instant by two parameters—amplitude and phase.* We have also seen in Unit 2 that many radiation detectors, and *all* optical detectors, respond to neither of these two quantities, but are instead sensitive to the time average of the *intensity*, that is to $\langle \psi^2 \rangle$, where ψ represents the disturbance of the wave as a function of time and space, and the angular brackets denote a time-average over several cycles of the wave. You will recall from Units 2 and 5, that $\langle \psi^2 \rangle$, the quantity measured by our detectors, is equal to $A^2/2$, where A is the wave amplitude. (See Fig. 18 also).

*That is, $\psi = A \cos \Phi$, where the instantaneous phase term Φ includes such parameters as wave velocity and frequency. I am neglecting polarization properties here.

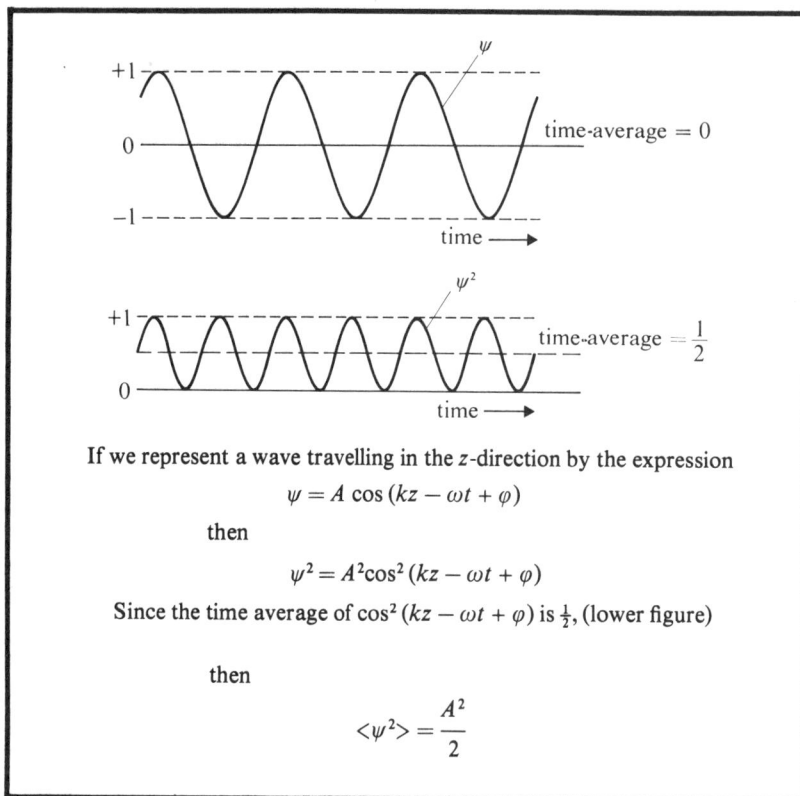

If we represent a wave travelling in the z-direction by the expression

$$\psi = A \cos{(kz - \omega t + \varphi)}$$

then

$$\psi^2 = A^2\cos^2{(kz - \omega t + \varphi)}$$

Since the time average of $\cos^2{(kz - \omega t + \varphi)}$ is $\frac{1}{2}$, (lower figure)

then

$$\langle\psi^2\rangle = \frac{A^2}{2}$$

Figure 18 An optical detector responds to the time-averaged intensity I, where $I = \langle\psi^2\rangle = A^2/2$.

We can see therefore, that all the possible optical detectors at our disposal throw away the phase information, leaving us with only the square of the amplitude. Hence, half the information about the waveform is lost whenever we try to detect the optical energy (for example with photographic film).

But we must be careful now, not to make the logically-false step that it is this lack of phase information on photographic film which is responsible for the loss of three-dimensional information. **It isn't.** We can see that it can't be as simple as this when we consider that the retina of our eye responds to the *time-averaged intensity* in just the same way as photographic film does. In other words, even in the live-viewing situation (never mind the recording situation) our eyes throw away the phase information. We are simply not sensitive to the phase information in a light wave. The very act of detecting the signal—of seeing—destroys that information. Yet you will find that books on holography stress the idea of storing the phase as well as the amplitude. Why? To answer this question we must return once again to our recording philosophy.

2.5 A different recording philosophy

When we explored the dummy-head recording technique, we found that decisions about head-movement and eye-focus were decided for us at the time the recording was made. We also came to the conclusion that this was not a particularly desirable state of affairs. So let us now try a different approach. This time, we shall try to avoid making irreversible decisions at the recording stage. We shall try to leave as many options as possible open to the viewer. This means, for instance, that all possible head-positions must be available on the recording, so that at the time of viewing you can choose whichever particular one you like—or even a succession of different ones. Similarly, it means that all possible eye-focus distances must be incorporated into the recording, so that once again you can choose whichever one, or whichever succession of different ones, you want at the time of viewing. This sounds like quite a tall order—to not only record the light intensity distribution patterns corresponding to the object viewed, but to record them from all possible viewing positions and with all possible depths of focus. Yet this is precisely what *holography* does. It does it, not by trying to simulate the human vision process when making the recording, but by storing the information in the form of the **optical field.** That is, not storing images, but storing information relating to the optical wavefronts. And to do this, both the amplitude and the phase of the waveforms are required.

holography

optical field

20

On playback we then regenerate the optical field—not images, but just replicas of the waves which were there at the time of recording. The formation of the image is then performed by our eyes *at the time of viewing*. The phase information is thrown away again in this viewing process, so our brains never make direct use of it. It was necessary to record it though, in order for us to store the optical field unambiguously and by so doing keep our viewing options open until the time of viewing.

3.0   Holography: The principles — *How to implement the philosophy*

3.1   Can we record phase?

Having decided on a recording philosophy we are now faced with the mammoth task of putting this philosophy into practice. I have, after all, just been emphasizing the fact that optical detectors respond not to amplitude and phase, but to time-averaged intensity. But since we have now decided that what we want to record is the optical field, how are we going to go about it? The amplitude is not too much of a problem. We have just seen that optical detectors respond to $A^2/2$, so by taking the positive square-root of the detector output we can deduce the amplitude. But what about phase? How do we record the phase information?

We were given a clue to the answer to this question in Units 7 and 8 when we asked the similar question, 'Can phase objects be made visible?' We saw in those Units that the answer was yes. Schlieren, dark-field and phase-contrast techniques all achieved this end. So we now ask, 'If phase objects can be made visible, can we use similar techniques to record the phase content of the propagating optical field?'

3.2   The Fourier transform encoding

To understand this question a little better, I want to quickly reiterate some of the fundamental ideas of the Fourier approach to imaging that we have adopted throughout this Course. Let's take it step-by step.

If any object (let's keep it two-dimensional for the moment) is illuminated with light, then the complex light-wave leaving the object carries away with it information about that object. The light field, immediately after interaction with the object, can be analysed in several ways.

One way is to divide the whole area of the object field into an array of points (with generalized coordinates (x_{ob}, y_{ob})) and assign relative amplitude and phase values to the field at each of these points.

An alternative approach (the Fourier approach) is to decompose the object field into a set of sinusoidal disturbances, each component of the set having its own *spatial* frequency. Each sinusoidal disturbance can then be assigned relative amplitude and phase values.

These two ways of expressing the optical field (immediately after it leaves the object) are equivalent. We can write—

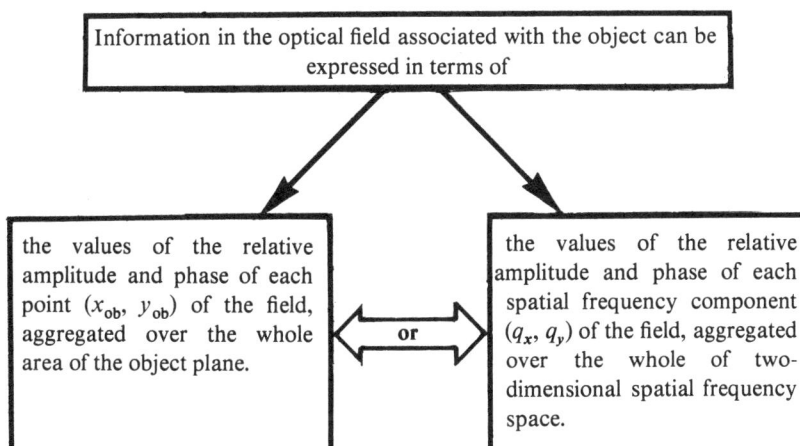

> Information in the optical field associated with the object can be expressed in terms of

the values of the relative amplitude and phase of each point (x_{ob}, y_{ob}) of the field, aggregated over the whole area of the object plane.	**or**	the values of the relative amplitude and phase of each spatial frequency component (q_x, q_y) of the field, aggregated over the whole of two-dimensional spatial frequency space.

But in an imaging system, there is also a plane (the back focal plane of the first lens) in which the object field's frequency components are directly transformed into position coordinates. That is, if the object is in the front focal plane of the lens (Fig. 19), then the distribution of amplitude and phase of the optical disturbance at the

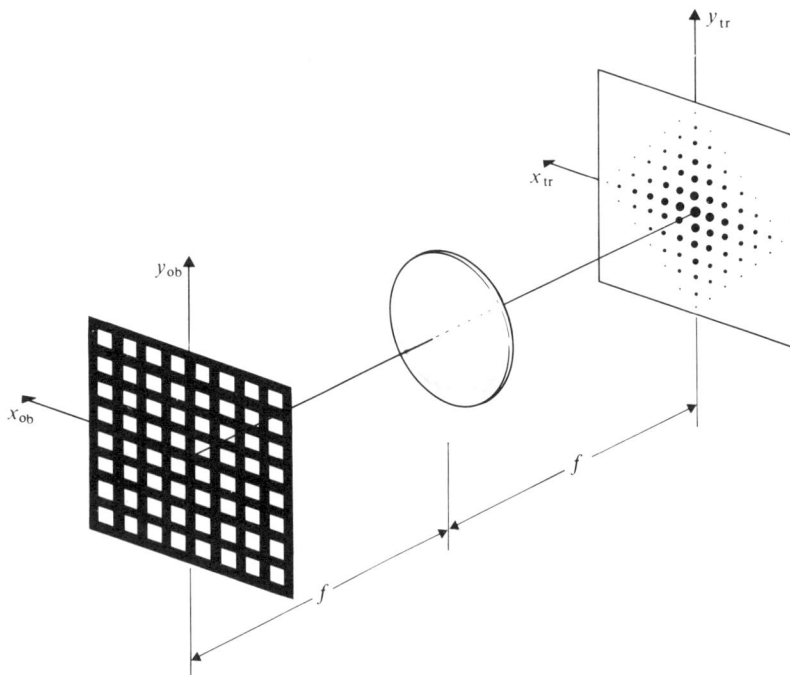

Figure 19 When an object is placed in the front focal plane of a lens, then the point-by-point field distribution in the back focal plane (as a function of x_{tr} and y_{tr}) can be interpreted as the Fourier transform of the object field.

object (as a function of the object coordinates, x_{ob} and y_{ob}) is related, *through the Fourier transform,* to the distribution of amplitude and phase of the optical disturbance at the back focal plane of the lens (as a function of the transform coordinates, x_{tr} and y_{tr}). The point-by-point distribution of amplitude and phase at the object has become a different point-by-point distribution of amplitude and phase in the back focal plane. What is more, if the coordinates x_{tr} and y_{tr} are replaced by the coordinates q_x and q_y, then this back focal plane point by point distribution can be interpreted as a direct plot of the object-field information expressed in spatial-frequency terms. We can write—

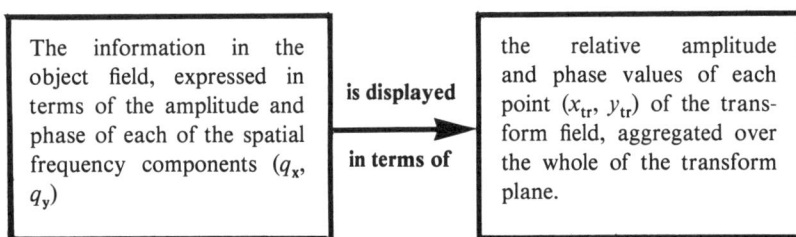

The information in the object field, expressed in terms of the amplitude and phase of each of the spatial frequency components (q_x, q_y)	**is displayed** **in terms of**	the relative amplitude and phase values of each point (x_{tr}, y_{tr}) of the transform field, aggregated over the whole of the transform plane.

So, given a perfect imaging system, the information (about an object) contained in the optical field at the transform plane, is the same information as is contained in the optical field at the object plane; *it is just encoded differently.* The information has been 'scrambled' as the wave has progressed down the imaging system.

Once familiar with the particular (i.e. Fourier transform) encoding used, it soon becomes possible to extract the information directly from the encoded form; it's just a question of learning the language. That is not to say that if I changed the code you would still be able to decipher the information. How could I change the code? Simple—I just look at a plane other than the Fourier transform plane. In fact, between the object plane and the FT plane there are an infinite number of other planes with a correspondingly infinite variety of codings. We shall see later that these other codings (which we shall call *Fresnel transform* encodings) are very im-portant in holography. The reason why we have especially emphasized the Fourier transform representation up until now is two-fold.

Fresnel transform

1   The Fourier transform is a well-defined, and relatively easily-manipulated mathematical operation.

2   The Fourier transform coding emphasizes a property of the object which is not usually apparent in the object plane itself—namely the object field's spatial-frequency distribution. A knowledge of this frequency distribution is very useful, since it allows us to modify the visual information in a frequency-selective way (as with spatial filtering, for instance), and perhaps by so doing, make the content more intelligible to human beings.

3.3   Phase and amplitude — a second stab

So, as far as holography is concerned we can record the amplitude and phase information in any plane. Whichever one we choose will contain all the information we need; only the coding will be different. But we haven't yet said how we are going to record this phase and amplitude information with detectors which are only amplitude-sensitive (or more strictly, sensitive only to $A^2/2$). Could it be, that one of those possible encodings we have just been talking about, would allow us to convert all the phase information in an object field into amplitude-only information? Well, we saw in Units 7 and 8 that a phase grating (i.e. an object which has no amplitude variations in it) produces an amplitude distribution in the Fourier transform plane. Perhaps this means that the Fourier transform plane is that plane where the information from a 'pure' phase object is converted into 'pure' amplitude information. If it is, then the obvious place to record a hologram is in this Fourier transform plane.

Is there any phase content to the optical field in the Fourier transform plane?

You already know, of course, that the answer to this question is that there is. But to make absolutely sure that you are convinced, let us consider recording the information in the Fourier transform plane photographically, and seeing whether or not our 'hologram' does contain all the information about the object. Consider the following 'thought' experiment.

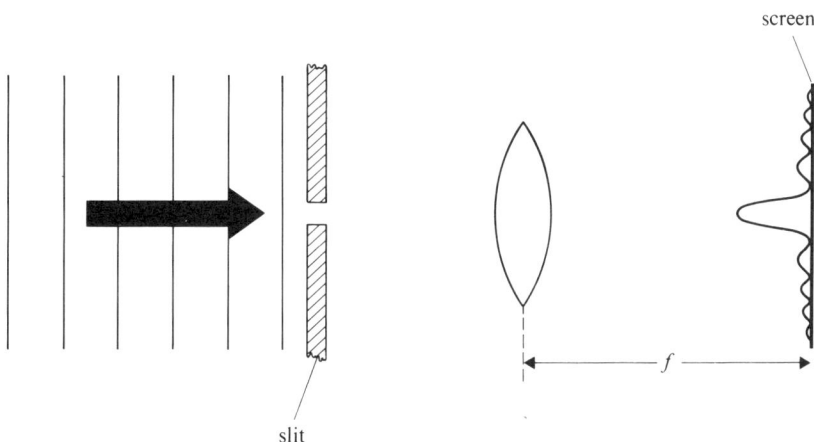

Figure 20   Producing the Fraunhofer diffraction pattern of a single slit.

Suppose we illuminate a rectangular slit with coherent light and produce its Fraunhofer diffraction pattern in the back focal plane of a lens (Fig. 20). We know that the amplitude distribution in the diffraction field has the functional form $\operatorname{sinc} q$,* since the sinc function is the Fourier transform of the slit function (Fig. 21(a)). If we were now to photograph the diffraction field, the film would not respond to the amplitude, but to the time-averaged intensity, i.e. to $A^2/2$. This function is shown on the right of Figure 21(b). Suppose we now develop the film in such a way that the amplitude transmittance of the transparency is directly proportional to the intensity distribution in the diffraction pattern.†

---

*Remember $\operatorname{sinc} q = \dfrac{\sin \pi q}{\pi q}$

†This can be done quite easily. It is required in holographic recording, and we shall discuss the details in Section 5.

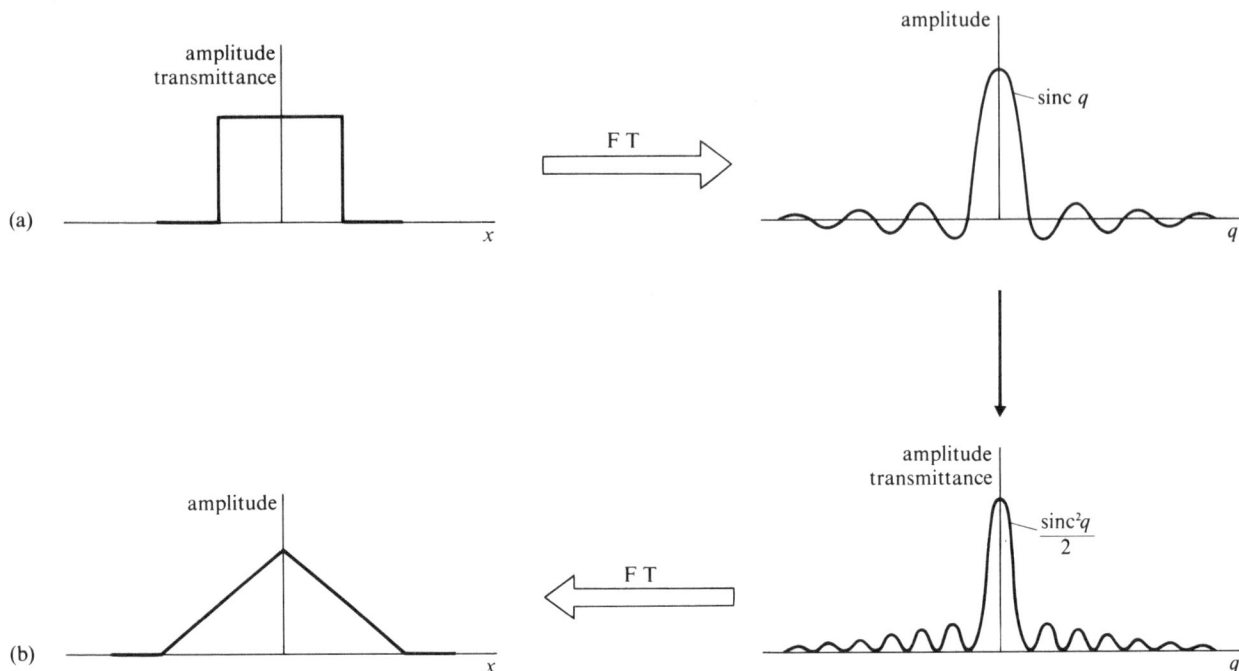

Figure 21 (a) The object field of a single slit and its Fraunhofer diffraction field.

(b) The consequence of photographing the Fourier transform distribution, and using it as an 'object' in a Fraunhofer diffraction experiment. The left-hand profile is the diffraction field which would result.

We now treat the transparency as a diffraction mask, illuminate it with coherent light, and look in the back focal plane of the lens for the diffraction pattern.

> Do you think this system will reproduce the image of the slit? Try explaining why, or why not, before reading on.

The answer is no. The diffraction mask has an amplitude transmittance proportional to $sinc^2q$. The Fourier transform of $sinc^2q$ is the triangular function

$$\text{and} \quad \left. \begin{array}{l} (a - bx) \text{ for } x > 0 \\[2mm] (a + bx) \text{ for } x < 0 \end{array} \right\} \tag{1}$$

shown on the left-hand side of Figure 21(b). Consequently, the *amplitude* distribution in the diffraction pattern will have this form, but the observed intensity distribution will have the form

$$\text{and} \quad \left. \begin{array}{l} (a - bx)^2 \text{ for } x > 0 \\[2mm] (a + bx)^2 \text{ for } x < 0 \end{array} \right\} \tag{2}$$

This function is sketched roughly in Figure 22.

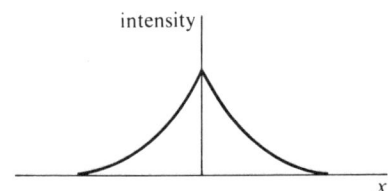

Figure 22

Obviously, the image doesn't look very much like the object. Photographing the Fourier transform pattern does not produce a hologram. The reason why this system doesn't work, is because there is hidden phase information in the sinc function shown in Figure 21(a). Every time the amplitude of this function is shown going negative, it means in effect, that the phase has changed by 180 degrees. We cannot record this phase information; we can only record A^2, and this can only give us the magnitude of the amplitude.

So, the Fourier transform plane is not, after all, a special plane in which all phase information is recoded as amplitude-only information. In fact, there is *no* plane in any imaging system—no possible other encoding—where such a transformation occurs.

24

If asked to write down generalized rules about how the information in an optical field can be reshuffled as the field progresses down an imaging system, we could write:

1   *any kind* of disturbance—whether a phase-only, amplitude-only, or mixed amplitude and phase disturbance—can be recoded into an *amplitude* and/or *phase* disturbance.

Except that

2   a *phase-only* disturbance can **never** be recoded as an *amplitude-only* disturbance

and

3   an *amplitude-only* disturbance can **never** be recoded as a *phase-only* disturbance.

> **SAQ 5**   (i) The optical field in the object plane of a given (perfect) imaging system, contains only *amplitude* information. Are there any other planes in this imaging system where the optical field can again be expressed purely in terms of amplitude information?
>
> (ii)   If the optical field in the object plane of the same imaging system, contains only *phase* information, is it possible this time to find other planes where the field can be re-expressed in terms of phase-only information?

3.4   The first Fourier transform hologram?

The German physicist, Fritz Zernike, was thinking about these problems in the mid-1940s, and in 1947 he performed an experiment which showed how the phase ambiguity in the experiment described earlier could be removed. His work attracted very little attention at the time, and even today it is frequently overlooked. Without realizing it, what he did in his experiment was demonstrate the principle of holographic recording, by making what today would be called a *Fourier transform hologram*. The strange thing was, that it never occurred to him to 'play back' the hologram and so reproduce an image of the original object. If he had done, it might have earned him his second Nobel prize!

Fourier transform hologram

Let me describe Zernike's experiment to you.

He illuminated a slit of width w with coherent light from a restricted-aperture, collimated arc-lamp, and then used a cylindrical lens to project the diffraction pattern of the slit onto a screen in the back focal plane of the lens. He observed the usual $\text{sinc}^2 q$ distribution (Fig. 25(a)). He then took a piece of plane-parallel glass and gave it a coating of silver thick enough to reduce the transparency to about 50 per cent. A very narrow scratch (of width much less than w) was then cut into this plate, so removing the silver at this point. The glass plate was then attached to the wider slit (of width w) in such a way that the line of the narrow scratch on the glass was aligned with the centre line of this wider slit. That sounds a bit convoluted, but I think Figure 23 makes the position clear. When this compound slit is illuminated,

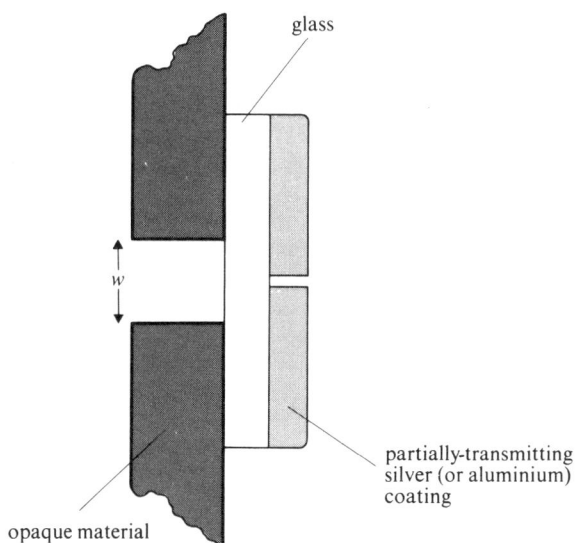

glass

w

partially-transmitting silver (or aluminium) coating

opaque material

Figure 23   Zernike used a compound slit to superimpose a coherent background onto a single slit diffraction pattern. By varying the thickness of the partially transmitting silver coating he could alter the phase relationship between these two components.

the very broad central diffraction lobe produced by the narrow scratch is *coherently superimposed* on the diffraction pattern from the slit of width w (Fig. 24). This is possible, because light from this wide slit can still reach the viewing screen (albeit slightly attenuated) by passing through the semi-transparent glass plate.

coherent superposition

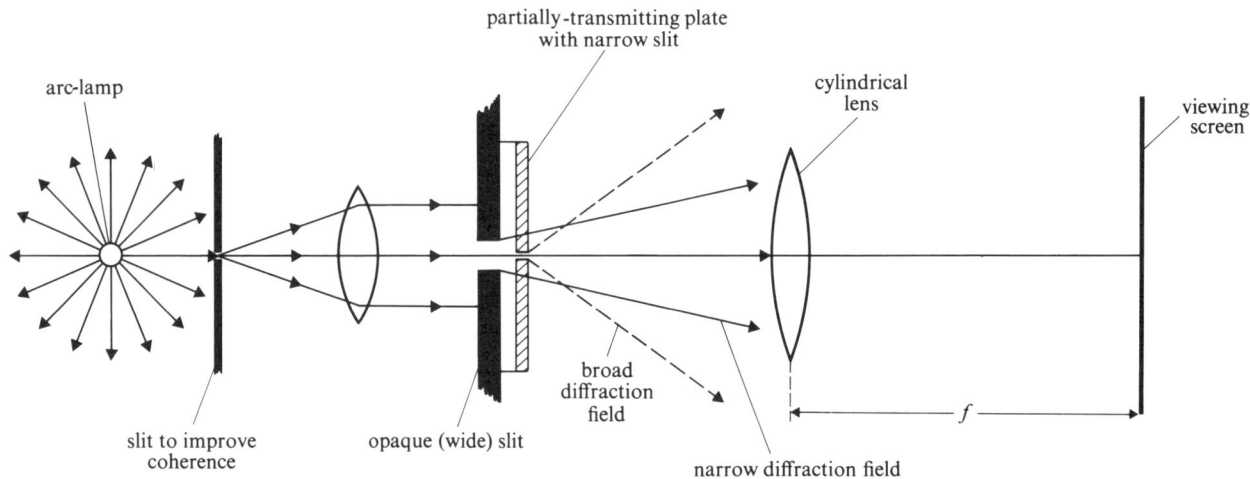

Figure 24 Zernike's experiment: the optical layout.

For the results shown in Figure 25, Zernike chose to use a glass plate in which the thickness of the silver coating was such as to make the coherent background (i.e. the central lobe from the narrow scratch) approximately antiphase with the central diffraction lobe obtained from the wide slit.* The distribution in Figure 25(b) shows the effect of superimposing these two diffracted waves. Notice that the spatial frequency of the fringes in this case (except near the centre) is *half* that found in the diffraction pattern from the single slit alone.

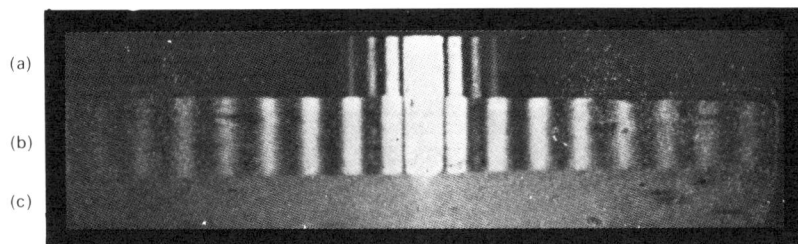

Figure 25 Zernike's experimental results: the Fraunhofer diffraction pattern of (a) the single slit alone; (b) the single slit with coherent antiphase background; and (c) the coherent background alone.

By considering Figure 26 we can understand how this change of spatial frequency comes about. The left-hand side of this Figure shows the single-slit-diffraction amplitude distribution (a sinc function), together with the observed, intensity distribution (a sinc-squared function). Notice that the squaring operation turns both negative and positive peaks into *maxima*. This effectively doubles the frequency of the fringes. If we now add the antiphase background to the diffracted light from the single slit, then, because this background is coherent with the single slit light, we must add the *amplitudes*. This shifts the whole of the combined amplitude function negative by an amount equal to the amplitude of the background. This shifted function is shown in the top right of Figure 26. When this function is squared, the spatial frequency of the single-slit *amplitude* distribution (the sinc function) is preserved, except near the central peak where there still remains some phase ambiguity.

How can we remove this ambiguity completely?

*Different silver thicknesses would produce different 'optical thicknesses', and hence different phase-relationships between the two diffracted waves.

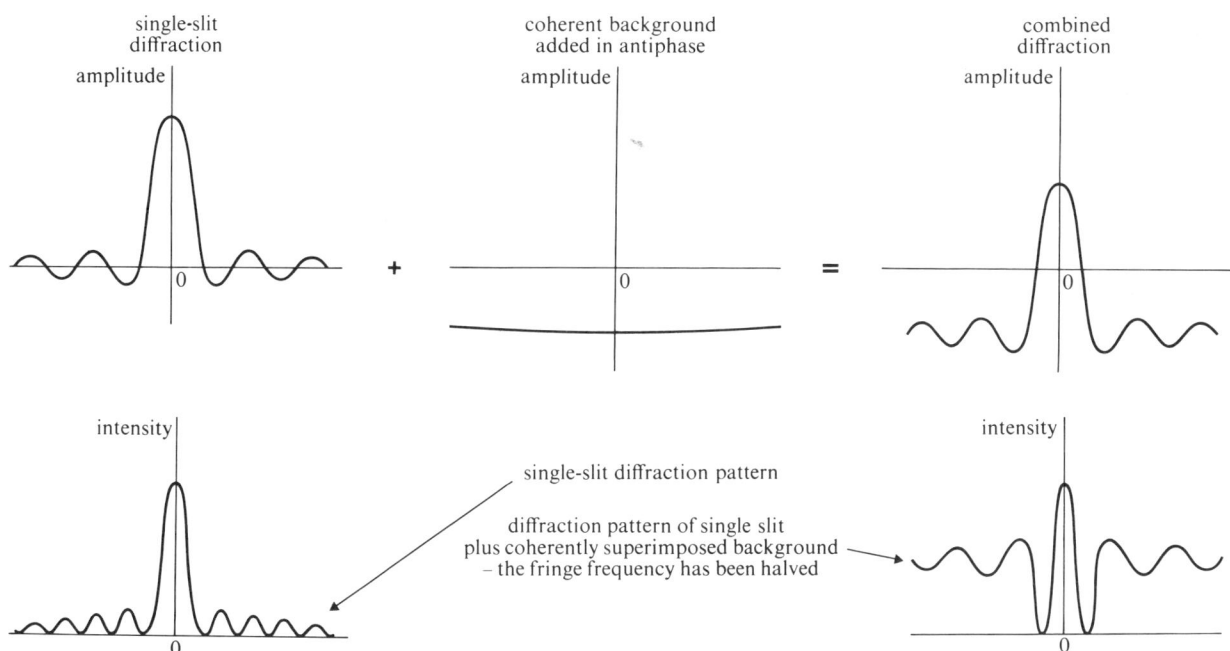

Figure 26   A graphical explanation of the photographs in Figure 25.

To completely remove this ambiguity, we must ensure that the amplitude of the coherent background is at least equal to the largest amplitude component in the single-slit diffraction wave, at all instants of time. In the example shown in Figure 26, we must shift the amplitude function sufficiently negative, for the combined function to lie completely below the 'Amplitude = 0' line.

SAQ 6   Is there any other way of removing all the phase ambiguity in the diffraction field; one which does not require the addition of such a large coherent background?

Although, at the present moment, Zernike's experiment might seem to be a long way from practical holography, it does in fact contain some of the principal notions involved in the subject. Let me list them.

1   In order to preserve phase information in the visual recording of an object, we have to add a coherent background signal.

2   The coherent background should be *at least* as large in amplitude as the largest amplitude component in the diffraction pattern of the object.

3   The phase and amplitude information is then preserved by the interference of the coherent background with the diffraction pattern. (In fact the interference can *only* occur if the background and object light are coherent.)

In addition, we have also seen from our discussion, that the Fourier transform plane is not a special plane in which all phase information is automatically converted into amplitude information. We could, in principle, select any other encoding (corresponding to a diffraction plane which is *not* the Fourier transform plane) and still record all the phase and amplitude information by adding in a coherent background.

4.0   Holography: Understanding the mechanism — *How to apply the principles*

4.1   Offset reference?

Zernike was limited in his experiments by the low coherence properties of available light sources, and in retrospect his method of obtaining a coherent background seems a great feat of ingenuity. Denis Gabor, when he made and 'played back' the first hologram in 1948, was similarly disadvantaged. In order to achieve the necessary spatial coherence of the source, he had to place a fine pinhole immediately in front of the source. Naturally this reduced the available intensity. (Indeed it reduced it to such an extent that only a system in which the diffraction pattern was coaxial with the background light could be considered.)

But with the advent of the laser in 1960, the problem of finding a light source with suitable coherence properties suddenly ceased to exist. Consequently, people began to experiment with other ways of adding in the coherent background. In particular, Leith and Upatnieks, working at the University of Michigan,* suggested adding in the coherent background at an angle to the main object light. We shall see that it was this new idea which dramatically ushered holography into the commercial arena, and hence into the public eye.

What happens if we add in the background light at an angle to the object light? This question is not, in fact, very easy to answer immediately. So let me take the problem in stages. Let me put to you a somewhat simpler question.

What happens when we add two coherently-related plane wavefronts together at an angle θ?

You should be able to answer this question (at least qualitatively), by now; we have asked very similar questions in several of the previous Units. Think about it for a while before reading on.

In fact, you have already met a piece of optical hardware which might well have been tailor-made for observing the overlapping of two plane wavefronts which are inclined at a slight angle to each other. Do you recall the Michelson interferometer which was discussed in Unit 5 and the corresponding TV programme? A schematic diagram of the optical layout of this instrument is shown in Figure 27. The light which reaches the viewing screen will have travelled either over the path ABCBE, or over the path ABDBE (or their dashed counterparts). If the beam-splitter is inclined at an angle of 45 degrees to the input beam, and if the two mirrors, CC' and DD',

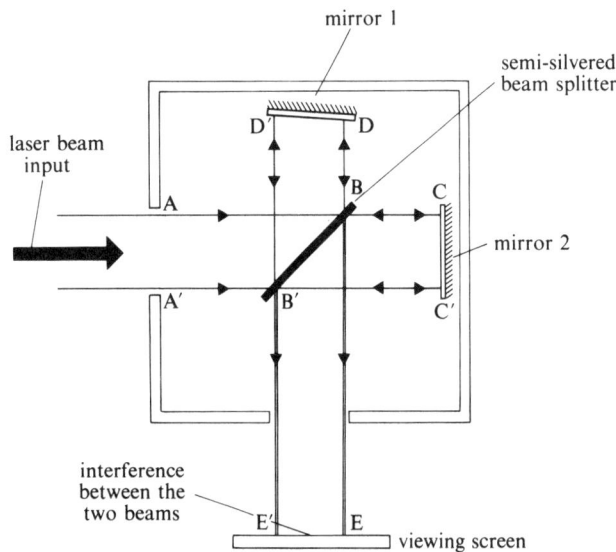

Figure 27 The optical layout of a Michelson interferometer.

---

*Leith and Upatnieks were working on the use of radar in communications engineering at the time. We can now see that the similarity between their radar work, and the wavefront reconstruction work of Gabor, exists because of the fact that a Fourier transform formalism is common to both fields of study. In the communications field, time is transformed into temporal frequencies, whereas in holography, space is transformed into spatial frequencies. This similarity between the two subjects has been exploited over and over again, often providing insights in areas where a knowledge of either field by itself would have led to a dead-end. This fact is exemplified by the large number of technical (i.e. jargon) terms which have been borrowed from the one field and used in the other. Terms such as 'side-band' and 'carrier frequency' for example, which are normally associated with the radio-broadcasting industry, are commonly to be found in texts on holography. We shall see the significance of these terms later on in these Units.

are both normal to the beam, then the angle of inclination between the two combining plane wavefronts at the viewing screen EE′ will be 0 degrees. Hence the optical field in the region EE′ will be determined by the phase relationship between the two waves arriving at the screen. If the optical path difference between the two routes ABCBE and ABDBE, is equal to a whole number of wavelengths then the two constituent waves arrive at EE′ *in phase*. There will be constructive interference (I am assuming that the path difference is much less than the coherence length of the illumination) and the whole field of view will be bright. If the optical path difference is increased (by racking back the mirror DD′ for example, while still keeping its surface normal to the beam), then there comes a point—when the optical path difference is equal to an odd number of half wavelengths—where the two waves arrive at EE′ *in antiphase*. They interfere destructively, and the whole field of view is dark. As the mirror DD′ is continuously moved back, the field of view changes cyclically from bright to dark to bright again, as the two component waves progressively change their phase relationship with each other.

In Figure 27 though, we have shown mirror DD′ *tilted at an angle* to the incident beam. This has the effect of causing one wavefront to arrive at EE′ at a small angle of inclination to the other wavefront.

What will be seen in the region EE′ now?

Now, the path difference between the two beams varies across the field from E to E′. The path difference [ABDBE − ABCBE] is *less* than the path difference [A′B′D′B′E′ − A′B′C′B′E′]. Between these two extremes there is a whole range of intermediate path differences. The field of view will therefore be crossed by a series of sinusoidal fringes, as shown in Figure 28.

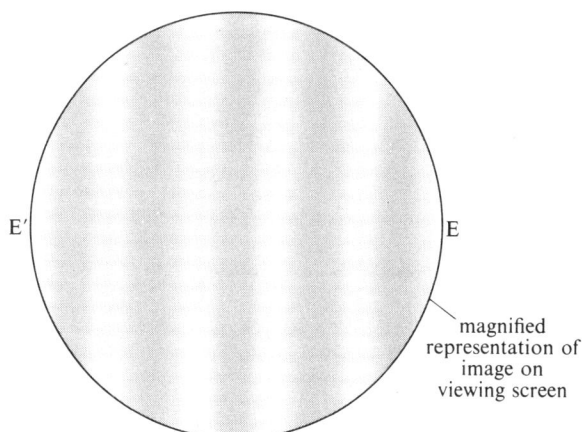

magnified
representation of
image on
viewing screen

Figure 28   The intensity distribution at the output of a Michelson interferometer when one of the mirrors is inclined at a small angle to the incident beam.

SAQ 7   Refer to Figure 28. Suppose that both E and E′ correspond to places of maximum *constructive* interference (i.e. the wavefronts are exactly a whole number of wavelengths out of phase at these points). By looking at the fringes shown in Figure 28, calculate how far D′ is behind D. (Disregard the fact that there may be imperfections in the optical components in the interferometer itself.)

So, now that we have a qualitative feel for what happens when two plane wavefronts interfere at an angle, we can look at this problem again—but a bit more analytically this time.

4.2   The superposition of plane wavefronts

Consider the diagram in Figure 29. Two coherent, equal amplitude, plane wavefronts are arriving at the plane BCDE with an angular separation of θ. The Figure shows a moment in time when wave ① has a maximum at the plane BCDE. Wave ② also is a maximum at this plane at the points B and D. At the points C and E wave ② is at a minimum. At points midway between B and C, C and D, and D and E, wave ② has zero displacement. In other words, the displacement value of

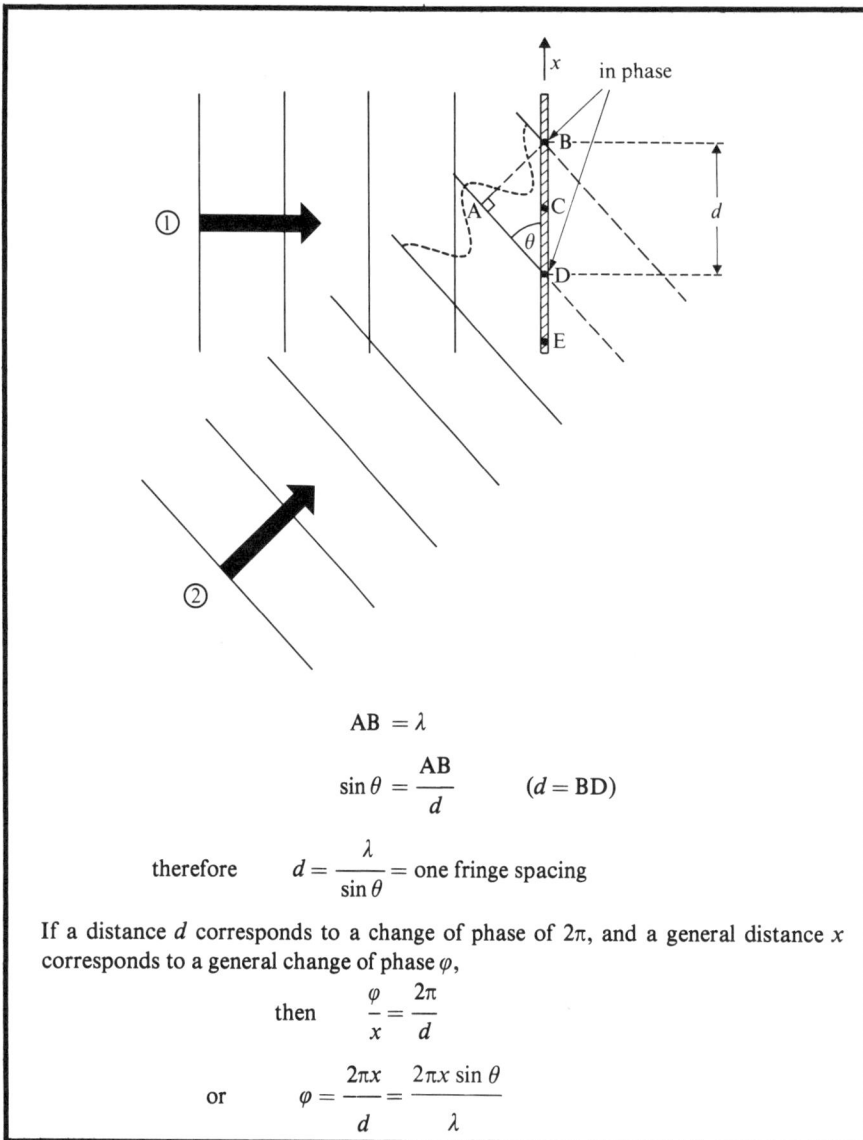

$$AB = \lambda$$

$$\sin \theta = \frac{AB}{d} \qquad (d = BD)$$

therefore $\quad d = \dfrac{\lambda}{\sin \theta} = \text{one fringe spacing}$

If a distance d corresponds to a change of phase of 2π, and a general distance x corresponds to a general change of phase φ,

$$\text{then} \qquad \frac{\varphi}{x} = \frac{2\pi}{d}$$

$$\text{or} \qquad \varphi = \frac{2\pi x}{d} = \frac{2\pi x \sin \theta}{\lambda}$$

Figure 29 Two plane waves, one incident normally, and the other incident at an angle θ to a photographic plate, generate static positions of zero and maximum field across the plate.

wave ② in the plane BCDE varies from point to point on this plane, whereas the displacement value for wave ① is the same at all values of x. We can say that there is a *phase* difference between waves ① and ② at the plane BCDE *which is a function of x*. We can find this function, $\varphi (x)$, quite easily.

$$\text{From the Figure} \qquad \frac{AB}{BD} = \sin \theta \qquad (3)$$

But $AB = \lambda$ (the wavelength of the light), and we can set $BD = d$ (the distance between two neighbouring points where the waves arrive in phase).

$$\text{Hence} \qquad d = \frac{\lambda}{\sin \theta} \qquad (4)$$

If the two waves ① and ② are arriving at B in phase, and then again at D in phase, it must follow that a whole 2π cycle of phase differences is encompassed between these two points. That is, going along x a specific distance d (from B to D) involves a change of phase difference between waves ① and ② from 0 to 2π

Consequently, a *general* movement of distance x along BCDE, must entail a change of phase difference from 0 to $2\pi x/d$.

But we have shown that

$$d = \lambda/\sin \theta$$

So the phase difference between waves ① and ② at points separated by a distance x is given by

$$\varphi (x) = (2\pi x \sin \theta)/\lambda \qquad (5)$$

Figure 29 shows the situation at only one instant of time. As waves ① and ② progress, we must continue to add together their instantaneous contributions at all points along the plane BCDE. Perhaps you can see from Figure 29 that these contributions from waves ① and ② will always be *in phase* at points such as B and D—a maximum displacement on wave ① will always coincide with a maximum on wave ②, a zero displacement on ① with a zero displacement on ② and a minimum on ① with a minimum on ②. At points like C and E, however, the contributions from the two waves will always be π out of phase (i.e. *antiphase*). These points will correspond to dark regions along the plane BCDE.

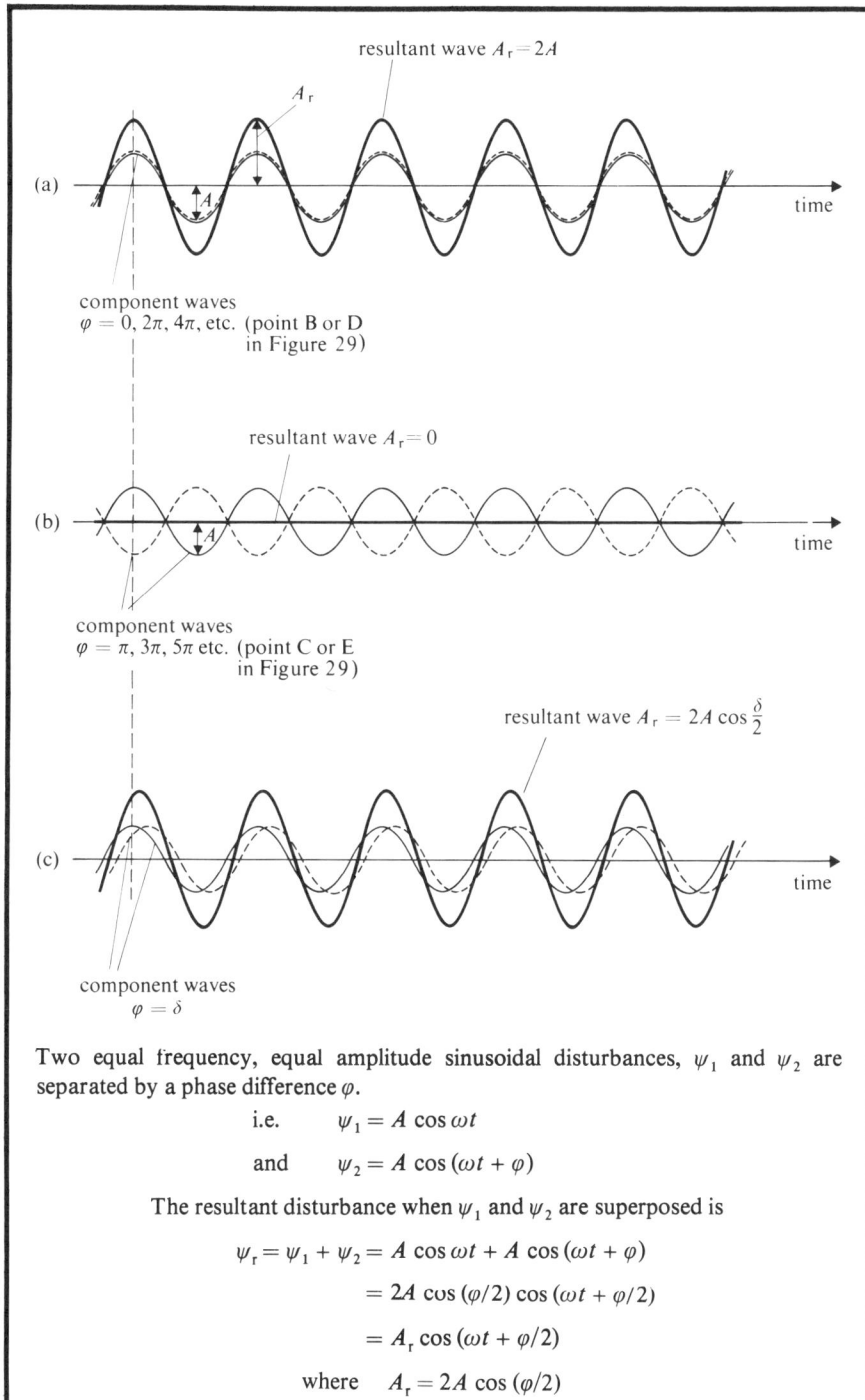

(a)   resultant wave $A_r = 2A$

A_r

A

time

component waves
$\varphi = 0, 2\pi, 4\pi$, etc. (point B or D in Figure 29)

(b)   resultant wave $A_r = 0$

A

time

component waves
$\varphi = \pi, 3\pi, 5\pi$ etc. (point C or E in Figure 29)

(c)   resultant wave $A_r = 2A \cos \frac{\delta}{2}$

time

component waves
$\varphi = \delta$

Two equal frequency, equal amplitude sinusoidal disturbances, ψ_1 and ψ_2 are separated by a phase difference φ.

i.e. $\psi_1 = A \cos \omega t$

and $\psi_2 = A \cos (\omega t + \varphi)$

The resultant disturbance when ψ_1 and ψ_2 are superposed is

$$\psi_r = \psi_1 + \psi_2 = A \cos \omega t + A \cos (\omega t + \varphi)$$
$$= 2A \cos (\varphi/2) \cos (\omega t + \varphi/2)$$
$$= A_r \cos (\omega t + \varphi/2)$$

where $A_r = 2A \cos (\varphi/2)$

Figure 30 Superposing two sine waves. Remember that the sum of two cosine terms can be expanded as

$$\cos X + \cos Y = 2 \cos \left(\frac{X + Y}{2}\right) \cos \left(\frac{X - Y}{2}\right)$$

(see page 13 of *Waves and Rays*).

Figure 30 shows how the contributions from our two equal-frequency, equal-amplitude, sinusoidal waves ① and ②, add together (as a function of time)

(i)   at points such as B and D, where the waves are in phase,

(ii)   at points such as C and E, where the waves are in antiphase, and

(iii)   at some intermediate point on the plane BCDE, where there is a phase difference $\varphi = \delta$ between the two waves.

From Figure 30, you can see that the maximum displacement of the *resultant* wave depends on the phase relationship between the two components, and has a magnitude

$$A_r = 2A \cos\left(\frac{\varphi}{2}\right) \tag{6}$$

where A is the amplitude of the component waves. Of course, this is simply the *maximum* excursion the resultant wave displacement can have. The displacement itself will oscillate back and forth between $\pm 2A \cos(\varphi/2)$ with a frequency v (which is the frequency of the light, and hence equal to c/λ).

If you found the kind of 'field-addition' diagrams that we drew in Units 3 and 4 helpful there, then you may also find a similar kind of Figure helpful in this situation. If you think you will, then you should now refer to Figure 31, and the text associated with it.

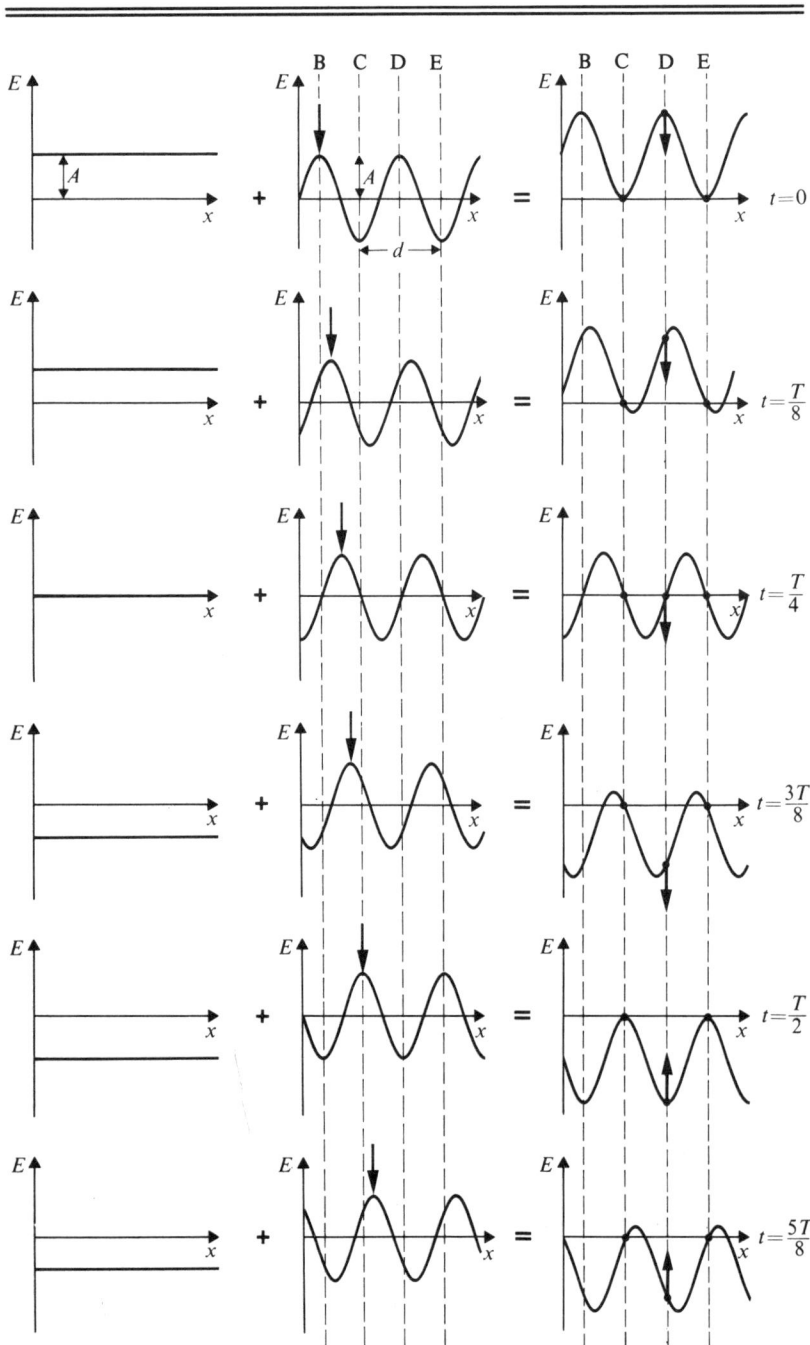

Figure 31 Wave ① is uniform across x (first column). Wave ② strikes the plane of the hologram obliquely. Therefore, at any instant there is a sinusoidal (sliced-off) field distribution across x. As time progresses, wave ② moves forward, and this sinusoidal distribution moves across x (second column). The third column shows the instantaneous resultant of waves ① and ② at successive times.

Figure 29 shows the position of the two-component wavefronts at one instant of time. Figure 30 shows how the optical field varies as a function of time at one or two discrete points across the plane BCDE. It is possible to partially combine these two Figures, by drawing out the displacement of the optical field at all points on the plane BCDE (i.e. as a function of x) for a few discrete instants of time within one period. I have done this in Figure 31 for the times $t = 0$, $T/8$, $T/4$, $3T/8$, $T/2$ and $5T/8$ where T is the time period of the optical oscillation, and is equal to $1/v$. In addition to showing the instantaneous resultant field in the plane BCDE, I have also shown the corresponding instantaneous states of the fields associated with the component waves ① and ②. You saw diagrams like these in Units 3 and 4. You should convince yourselves that Figures 29, 30 and 31 are self-consistent. In particular you should notice how both the component fields are in phase at points B and D, and in antiphase at points C and E. You should also notice that points such as C and E in the resultant field have zero displacement at all times during the cycle, whereas points B and D in the resultant field oscillate between $+2A$ and $-2A$. Intermediate points oscillate over a smaller range of displacement; they also do not oscillate in synchronism with points such as B and D.

To calculate the extent of the optical field's excursion at *any* point on the plane BCDE, we must combine equation 5 with equation 6. For if the resultant amplitude generated by the two interfering waves depends on the phase difference between the waves (equation 6), and if the phase difference between the waves varies as we move across x (equation 5), then it is clear that the resultant amplitude will vary across x as

$$A_r(x) = 2A \cos \left[\frac{2\pi x \sin \theta}{2\lambda} \right] \tag{7}$$

where I have substituted equation 5 into equation 6.

4.3   Making a sinusoidal grating

We can now consider what happens if we put a photographic plate in the plane BCDE, and try to record this (co)sinusoidal variation in amplitude. As you know, the photographic plate can only respond to the time averaged intensity $I(x)$; that is to $A_r^2/2$.

So, from equation 7, we have

$$I(x) = 2A^2 \cos^2 \left[\frac{\pi x \sin \theta}{\lambda} \right]$$

$$= A^2 + A^2 \cos \left[\frac{2\pi x \sin \theta}{\lambda} \right] \tag{8}$$

where I have used the identity $\cos 2\theta = 2 \cos^2 \theta - 1$.

That is, we have a (co)sinusoidal intensity distribution across x, the spatial frequency of which depends upon θ and λ. In fact, since the spatial frequency q, is just the inverse of the spatial wavelength d, we can write

$$q = \frac{1}{d} = \frac{\sin \theta}{\lambda} \tag{9}$$

Hence the larger θ, or the smaller λ, the closer together will be the intensity maxima across BCDE.

If we can process the photographic plate in such a way that the *amplitude* transmittance of the developed plate is proportional to this intensity distribution,* then we will produce a (co)sinusoidal grating (like the ones you met in Units 3 and 4) with a spatial frequency of $\frac{\sin \theta}{\lambda}$. You can easily see from the expression for $I(x)$

*We shall see in Section 5.2 that it is not possible to completely meet this requirement for the particular intensity distribution described by equation 8. The problem is that the photographic emulsion is not linear in the region of $I(x) = 0$ (and in equation 8, $I(x)$ does equal zero for certain values of x). This difficulty is removed if the amplitudes of the two waves ① and ② are made unequal.

that the amplitude transmittance would then have the spatial-frequency spectrum shown in Figure 32.

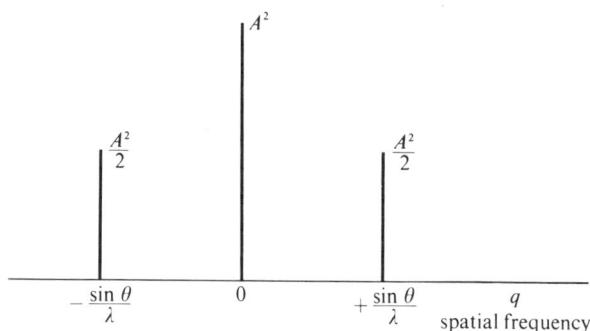

Figure 32

The effect of illuminating this grating with a parallel beam of coherent light should by now be well known to you. The three 'output' waves are shown in Figure 33.* Notice that each of the beams is angularly separated by θ, the same angle that existed between the two beams which generated the grating.

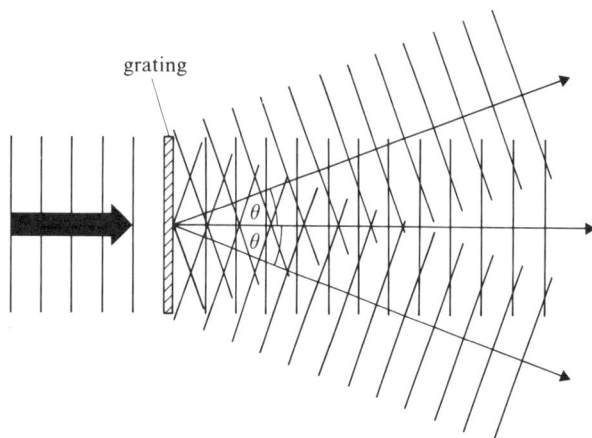

Figure 33 This hologram is a sinusoidal amplitude-grating

In the example I have just discussed, the two overlapping waves ① and ② were supposed to have equal amplitudes. Furthermore, wave ① was normal to the film plane BCDE (Fig. 29). These are two rather special conditions. In the more general case, when wave ① has amplitude A_1 and is incident on the plane BCDE at an angle θ_1, and wave ② has amplitude A_2 and is incident on the film plane at angle θ_2, the general result for the intensity distribution across x can be written as

$$I(x) = \frac{A_1^2}{2} + \frac{A_2^2}{2} + A_1 A_2 \cos\left[\frac{2\pi x}{\lambda}(\sin\theta_1 - \sin\theta_2)\right] \qquad (10)$$

where the angles θ_1 and θ_2 are measured in the same direction from the normal to the film plane.

> **SAQ 8** Two plane waves ($\lambda = 633$ nm) have an angle of θ between their directions of propagation. A photographic plate is positioned so that these waves strike the plate symmetrically (with respect to the normal), as shown in Figure 34. If $\theta = 50°$ calculate the spatial frequency of the recorded fringes.

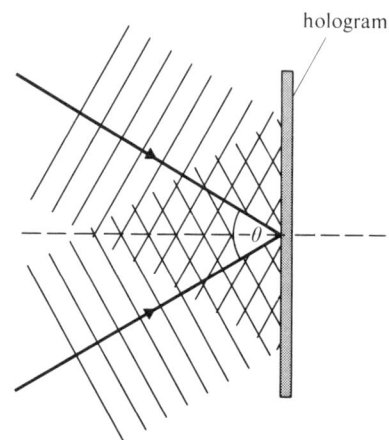

Figure 34

A grating with an amplitude transmittance proportional to the expression for $I(x)$ given in equation 10, would behave in essentially the same way as our previous sinusoidal grating; it will, on illumination, generate three beams as before. The most important difference arises from the 'unequalling' of A_1, and A_2. This causes a

---

*You will probably have noticed that in these Units I tend to use the terms 'parallel beam' and 'plane wave' interchangeably. This is not accurate of course, since you will remember from Units 3 and 4 that if an infinite plane wave is restricted, this very restriction will introduce other plane wave components travelling in slightly different directions from the main wave. This explains why a laser beam diverges—we just cannot produce truly parallel beams of coherent light! Nevertheless, throughout this discussion this is the approximation I am using (it is known as Abbé's approximation). In practical terms it means that I am neglecting diffraction caused by the finite width of the beam. As long as I bear this in mind (I can always consider the effect of this diffraction later), the Abbé approximation will always lead to results which are substantially correct.

bigger proportion of the energy to go into the 'straight-through' beam, as can be seen from the $I(x)$ spectrum shown in Figure 35 (where $A_1 = \frac{2}{3}A$ and $A_2 = \frac{4}{3}A$).

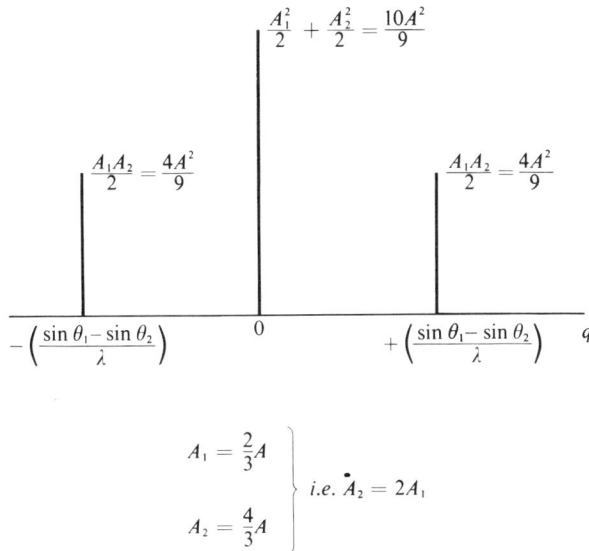

$$\frac{A_1^2}{2} + \frac{A_2^2}{2} = \frac{10A^2}{9}$$

$$\frac{A_1 A_2}{2} = \frac{4A^2}{9} \qquad \frac{A_1 A_2}{2} = \frac{4A^2}{9}$$

$$-\left(\frac{\sin\theta_1 - \sin\theta_2}{\lambda}\right) \qquad 0 \qquad +\left(\frac{\sin\theta_1 - \sin\theta_2}{\lambda}\right) \qquad q$$

$$\left.\begin{array}{l} A_1 = \frac{2}{3}A \\[2mm] A_2 = \frac{4}{3}A \end{array}\right\} \quad i.e.\ \overset{\bullet}{A}_2 = 2A_1$$

Figure 35

In case you found the mathematics in this Section tedious (possibly so much so that you skipped it!), let me summarize the results for you here—because the results *are* important.

1   If two coherent plane-waves intersect a photographic plate at angles θ_1 and θ_2, they generate on the plate a sinusoidal amplitude-grating of spatial frequency $(\sin\theta_1 - \sin\theta_2)/\lambda$.

2   This spatial frequency distribution can be thought of as *modulating* a uniform background level. The depth of the modulation (which we define as $(I_{max} - I_{min})/(I_{max} + I_{min})$) depends on the ratio of the amplitudes of the two plane waves. Maximum modulation is achieved when the two amplitudes are equal. (We shall go into the details of this in Section 5.2.)

3   If the grating is 'played back' with a plane wave of light of the same wavelength as was used in the recording, and incident in the same direction as one of the waves which generated the grating, then three output beams will be produced. One of these will continue in the same direction as the incident wave; the other two will emerge symmetrically on either side of this beam (Fig. 33). The angle between the 'straight-through' beam and either of the flanking beams is equal to the angle between the two waves, ① and ②, with which we produced the grating.

4   The output waves are plane waves—they 'focus' at infinity.

4.4   What about phase?

In the previous Section, I showed that the depth of modulation of the fringes on the photographic plate was related to the ratio of the *amplitudes* of the two waves. But how is the fringe structure modified if the *phase* surfaces of waves ① and ② are not identical? Suppose, for instance, that wave ② in Figure 29 had had a slightly distorted wavefront. How would the pattern of fringes be affected? Well, we don't need to go into the details here; all we need is a qualitative understanding of what would happen. And I think we can get this by examining the expression we derived for the phase variation across the plane BCDE (in Fig. 29) as a function of x, when wave ② was a plane wave.

This was $$\varphi(x) = \frac{2\pi x \sin\theta}{\lambda} \qquad \text{(Eq. 5)}$$

Hence for fixed values of θ and λ we see that φ is directly proportional to x. This was why the fringe structure, which is given by $\cos(\varphi/2)$, was (co)sinusoidal across x.

But if wave ② is not a plane wave, but is distorted in some way, then equation (5) will not be true. There will be no simple proportionality between φ and x, and hence the fringe structure will not be simply (co)sinusoidal. The distorted wavefront will produce a distorted phase relationship between φ and x, and this will distort the (co)sinusoidal fringe structure.

In general, we can say that phase information will be encoded into the *shape* of the interference fringes, and amplitude information into the *depth of modulation* of the fringes.

So let us see if we can understand how this works when the amplitude and phase information is generated by a real object.

4.5 A hologram of a sinusoidal amplitude-grating

In Fourier analysis terms, the simplest object we can think of is a sinusoidal amplitude-grating. What would happen to the interference fringe pattern if we inserted this grating into beam ② (Fig. 36)? We know how the input plane wave would be modified by the sinusoidal grating—it would produce three output beams with an angular distribution dependent on the spatial frequency of the grating. We must now ask how these three beams would interact with a coherent (reference) background. Just to keep the argument and analysis clear at this stage, let us suppose we position the grating sufficiently far away from the detecting photographic plate for the three beams to have separated into three distinct regions (Fig. 36). This essentially means that in the region of the plate we have something closely approximating to the Fraunhofer diffraction pattern of the grating. It is clear from Figure 36 that we now have three regions on the plate where there is overlap between reference and object beams—namely AB, CD and EF. The fringes in these regions will not be identical, since the phase angle for region AB is θ_1, for CD is θ_2 and for EF is θ_3 (where $\theta_1 > \theta_2 > \theta_3$).

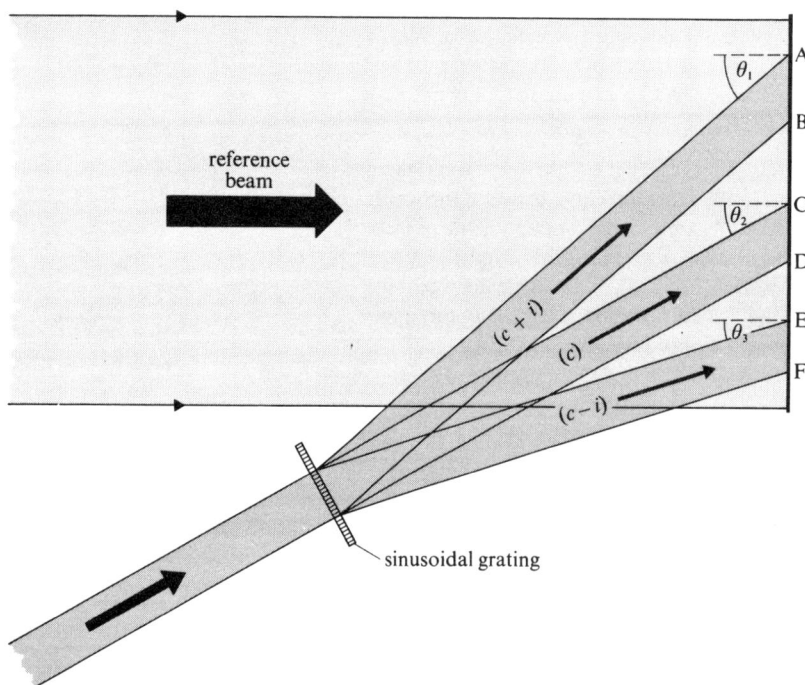

Figure 36 Making a hologram of a sinusoidal amplitude-grating object. Incidentally, 'c' stands for 'carrier' and 'i' for 'information' (see Section 4.8).

From Section 4.2, we know that the fringe spacing in each of these regions is given by the equation

$$d = \frac{\lambda}{\sin \theta} \qquad \text{(Eq. 4)}$$

and that the fringes will be sinusoidal. Hence, in the region AB, we have sinusoidal fringes of frequency

$$q_1 = \frac{1}{d_1} = \frac{\sin \theta_1}{\lambda} \tag{11}$$

in region CD, the fringe frequency is

$$q_2 = \frac{1}{d_2} = \frac{\sin \theta_2}{\lambda} \tag{12}$$

and in region EF, the frequency is

$$q_3 = \frac{1}{d_3} = \frac{\sin \theta_3}{\lambda} \tag{13}$$

The resultant pattern on the plate is shown schematically in Figure 37. This is our hologram. Remember that in reality the fringe structure would be slightly more complicated than this, because of the other (weaker amplitude) components in the diverging beams which would have to be considered when we took finite beam-width diffraction into account.

We must now consider what happens in the replay situation. Suppose we use the reference beam as the reconstruction beam (Fig. 38). Then each of the regions AB, CD, EF on the hologram will behave like a sinusoidal diffraction grating, and *each* will produce three emergent beams. The difference is that each 'grating' has a different spatial frequency, so that the angular separation of the beams from each of the gratings will be different. The finer the fringes, the greater is this angular separation.

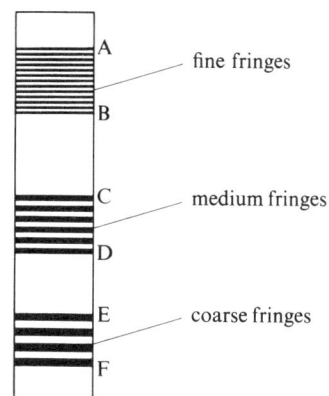

Figure 37   A stylized representation of the fringes on the hologram shown in Figure 36.

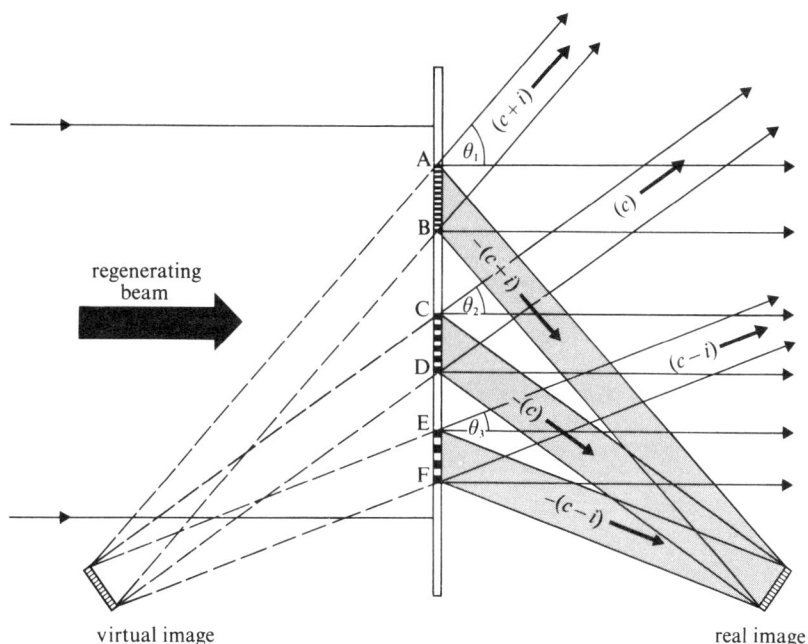

Figure 38   Replaying the hologram of the sinusoidal-grating object. The images are formed where all the spatial frequency components once again overlap.

Now in order to generate an image of the 'object' (sinusoidal grating) that we started with, it is necessary to make the three plane-wave components which make up the optical field associated with this grating (I labelled them c, $(c + i)$ and $(c - i)$ in Fig. 36) overlap again. I think you can see from Figure 38 that on replaying the hologram there are two distinct regions where this condition is satisfied.

A *real* image is formed in front of the hologram plate where the three-component beams really do intersect each other, and a *virtual* image is formed behind the plate in the region from which the other set of three-component beams appear to have diverged. These images are symmetrical with respect to the plane of the hologram.

In addition to this diffracted light distribution, a good deal of the reconstructing light beam will go straight through the hologram. In the regions where there was no (or at least very little) fringing (e.g. BC and DE), it will pass through virtually un-attenuated, whereas in the region of the gratings the light will be attenuated much more strongly.

4.6 More complex two-dimensional transmission objects

Having now seen that we can make a hologram of a simple sinusoidal-grating 'object', you should have no qualms about extending the argument to cover any other one- or two-dimensional objects. You saw in Units 3 and 4, that the optical field associated with more complex *one*- dimensional objects (say a square-wave grating, or a double slit) could be 'synthesized' by adding in further spatial-frequency components, each with the correct amplitude and phase values. These extra components in the optical field give rise to extra sets of plane waves, and these plane waves in turn give rise to additional sinusoidal fringes at the hologram plate.

A two-dimensional object can be dealt with in a similar way. The only difference now is that the two-dimensional object field will have spatial frequency components in both the x and y directions (i.e. q_x and q_y), and this will lead to a 'two-dimensional' superposition of sinusoidal fringes at the hologram. Even phase objects can be analysed in this way—as you should recall from Units 7 and 8.

I don't think you should now need to try too hard to convince yourselves that, on playback, each one of these sinusoidal fringe patterns on the hologram will work in a way similar to that depicted in Figure 38, and so produce a superposition of sinusoidal disturbances in the region of the real and virtual image. If no plane-wave components are lost or distorted in this process, then the sinusoidal components which are used to synthesize the image fields on playback, should be identical to those components into which we 'decomposed' the object field during the recording. Hence, the two image fields should be exact replicas of the original object field.

4.7 Three-dimensional objects

At least, such a statement is true for two-dimensional objects. But what if our object is three-dimensional? We can begin to consider this question by looking at the relationship between object and images as a function of the precise position of the two-dimensional object plane during recording. Look at Figure 39. The object is

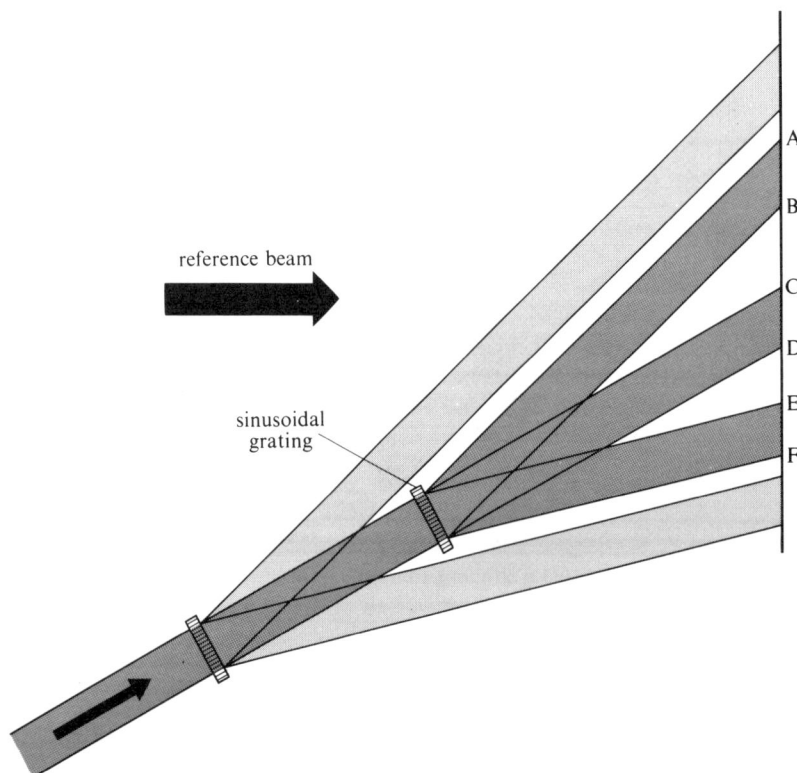

Figure 39 A movement in the position of the diffracting object leads to an equivalent movement in the position of the two reconstructed images.

once again the simple sinusoidal amplitude-grating. When this object is positioned close to the holographic plate, we get, as before, fringes in the regions AB, CD and EF. The frequency of the fringes depends on the angle at which the beam strikes the plate. If the sinusoidal-grating object is now moved away from the holographic plate to the other position shown in the Figure, then the diffraction at the object will be exactly the same as it was in the previous position. But now because the object is further away from the plate, the beams will have spread further apart before they intersect it. They will, however, strike the plate *at exactly the same angle* as they did previously. Consequently, the spatial frequency of the fringes will remain unchanged, but their location on the plate will be altered.

This fact will be reflected in the playback operation. The hologram will generate diffraction beams at the same angles as before, but because they originate from points on the plate further separated than previously, they will travel further before overlapping. Hence, as the object is moved towards the holographic plate, the images also move towards the plate. Similarly, as the object is moved to a plane further away from the hologram, so the image planes move away from the hologram by the same amount. The *virtual* image will be reconstructed in exactly the same plane that the object occupied during recording; the *real* image is mirror-symmetrical with this virtual image relative to the plane of the hologram.

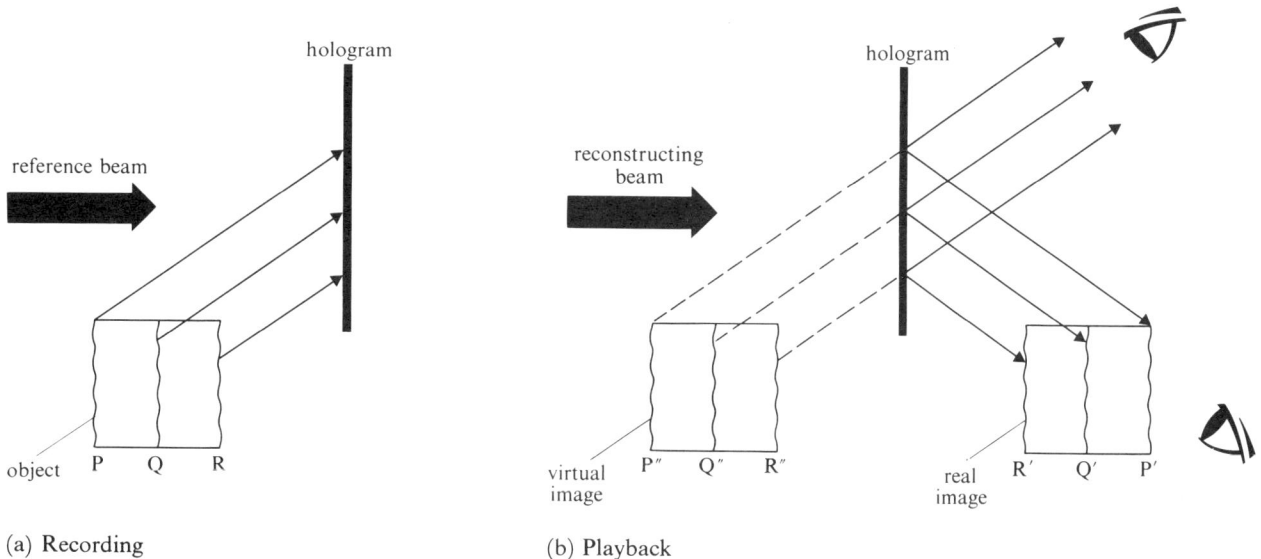

(a) Recording

(b) Playback

Consequently, if the object has three-dimensional depth, then the two images will also have the same three-dimensional depth; *but with one exception*. Figure 40 summarizes the recording and playback situations described above. Notice that, because of the mirror symmetry of the real and virtual images, the planes in the virtual image bear a one-to-one correspondence with the planes of the object (as viewed by the eye looking through the hologram), whereas in the real image, the order of the planes is reversed. Thus, although P is behind Q, and Q is behind R in the object, and similarly P″ is behind Q″, and Q″ is behind R″ in the virtual image, in the *real* image the eye sees P′ *in front* of Q′, and Q′ *in front* of R′, *even though* P′ *may be obscured by* Q′ *and* Q′ *obscured by* R′! This peculiar property of the real image (which has to be seen to be fully appreciated) is known as *pseudoscopy*.

Figure 40 The relationship between various planes in the object and those same planes in the hologram's real and virtual images. Notice the pseudoscopy ('back-to-frontness') of the real image.

pseudoscopy

> **SAQ 9** Figure 40 would suggest that some of the light leaving plane P on the three-dimensional object would interact with plane Q and be rediffracted. Similarly, some of the light from plane Q looks as though it could be rediffracted by plane R. Wouldn't this complicate the interference fringes on the hologram?

4.8   Side-band Fresnel holograms

When we set ourselves the problem of making a hologram of a sinusoidal grating (Section 4.5), we specified that the principal orders from the grating should have separated by the time they reached the holographic plate (Fig. 36). You can see from Figure 41, that this is equivalent to saying that

$$l \geqslant \frac{ad}{\lambda} \qquad (14)$$

For a grating with a spatial frequency of 100 lines per mm ($d = 10\ \mu m$), and an illuminating He-Ne laser beam of width 1 cm, equation 14 says that l must be greater than about 18 cm (see Fig. 41). If the spatial period of the grating, on the

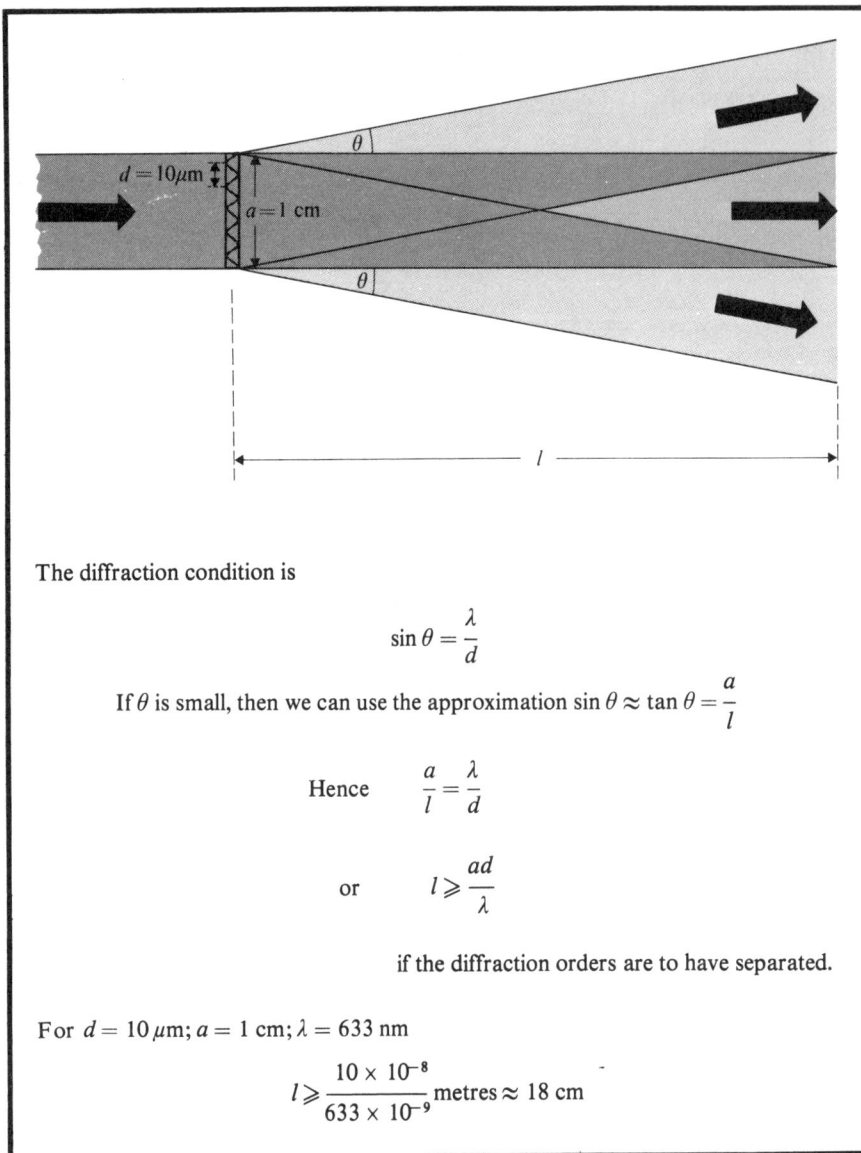

The diffraction condition is

$$\sin \theta = \frac{\lambda}{d}$$

If θ is small, then we can use the approximation $\sin \theta \approx \tan \theta = \frac{a}{l}$

Hence $\quad \frac{a}{l} = \frac{\lambda}{d}$

or $\quad l \geqslant \frac{ad}{\lambda}$

if the diffraction orders are to have separated.

For $d = 10\,\mu\text{m}; a = 1\,\text{cm}; \lambda = 633\,\text{nm}$

$$l \geqslant \frac{10 \times 10^{-8}}{633 \times 10^{-9}}\,\text{metres} \approx 18\,\text{cm}$$

Figure 41 A first approximation to the Fraunhofer diffraction condition.

other hand, was 1 mm, then l would have to be greater than about 18 metres. When you bear in mind that these calculations are not much more than conservative estimates—we have neglected the diffraction caused by restricting the width of the laser beam to 1 cm, for instance—then you will appreciate that l should be much greater than these calculated values if the photographic plate is to be placed anywhere near the *Fraunhofer* diffraction region. But in the case of the 1 mm period grating, where l should be much greater than 18 metres, this is obviously well nigh impossible with any practical holography set-up. Yet for every-day objects, a spatial frequency of one line per mm may well correspond to relatively *fine* detail—the coarser detail in the object would require l to be even greater than 18 metres! In other words, with the kind of distances typically used in a holography set-up, we don't normally have an earthly chance of being in the Fraunhofer diffraction region. The diffraction pattern which falls on the holographic plate will be the *Fresnel* (or near-field) diffraction pattern. Figure 42 shows (although still in an exaggerated way) how the principal orders of a sinusoidal-grating diffraction pattern would strike the photographic plate. You can see that in this situation, the different fringe-frequency regions will not occur at separate and distinct regions of the plate, but instead will overlap across most of it. This will make the fringe structure at the hologram much more complicated, but nevertheless the complicated structure should still be decomposable into its sinusoidal components, and reconstruction of the image should still proceed in exactly the same way as before.

There will be one difference though. In addition to the possibility of each of the diffraction orders interfering independently with the reference beam, we now have the extra possibility of the orders interfering with themselves (i.e. c with $(c + i)$ and c with $(c - i)$ for example).

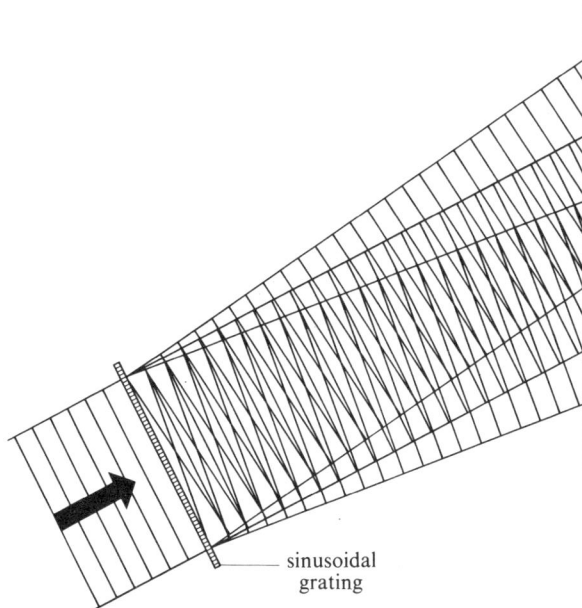

Figure 42 In the Fresnel diffraction region the spatial frequency components are still jumbled together.

sinusoidal grating

We would expect these fringes to generate extraneous images. There are two factors however, which stop these extra images from being troublesome. If the amplitude of the reference beam is made much larger than the total object-light amplitude falling on the film (a requirement which is also needed for other reasons in practical holography—see Section 5.2), then the 'self-interference' fringes will be much weaker than the fringes produced by interference with the reference beam. In addition, the angle between diffraction orders is likely to be much smaller than the angle between any individual diffraction order and the reference beam. Hence on replay, these very coarse 'self-interference' fringes will cause only slight diffraction divergence of the straight-through beam; this is not likely to interfere with the much more strongly-diffracted imaging beams. Apart from this, though, there appears to be no reason why the 'near-field' hologram shouldn't work in exactly the same way as the 'far-field' hologram. We have simply stored the information in one of those other codings I was talking about earlier.

When the hologram is made in this way (i.e. in the near-field of the object) we describe it as a *Fresnel hologram,* because the photographic plate is in the Fresnel diffraction region of the object. Furthermore, when the reference beam is folded-in at an angle to the zero-order object beam, we describe the resulting hologram as a *side-band Fresnel hologram.* This is the most common form of hologram made; certainly the hologram in your Home Kit, and the holograms you will make yourselves are of this kind.

Fresnel hologram

side-band Fresnel hologram

Why 'side-band'? Well let me explain by quickly summarizing the holographic mechanism.

1   Two beams intersecting at an angle θ generate sinusoidal interference fringes of frequency $(\sin \theta)/\lambda$ (assuming that the reference beam is normal to the detection plane). These fringes are shown in Figure 43(a). There is no object in either of the beams, so the fringes carry no information about this 'non-existent' object. By analogy with radio transmission, these fringes are often referred to as *carrier fringes.* *

carrier fringes

2   If a one-dimensional object (a sewing-needle is near enough to this ideal, for example) is placed in one of the beams, then some of the light in this beam will be diffracted. This diffracted light will intersect the reference beam at angles slightly above, and slightly below the zeroth order angle defined by θ in (1) above. Consequently, the spatial frequency of the sinusoidal carrier-fringes will be slightly increased and decreased by the diffracted light. We could say that the carrier-frequency fringes have been modulated by the diffraction information from the sewing-needle. Figure 43(b) shows this degree of modulation. The amount of deviation from the carrier frequency depends on the diffraction properties of the

*The carrier wave in radio is the one you 'tune-in' to (e.g. 1 500 metres for Radio 2, longwave). It contains no information itself, as you can prove to yourself by tuning in at 4.00 a.m. in the morning! Information is impressed onto this carrier wave by modulating either its amplitude (AM) or its temporal frequency (FM).

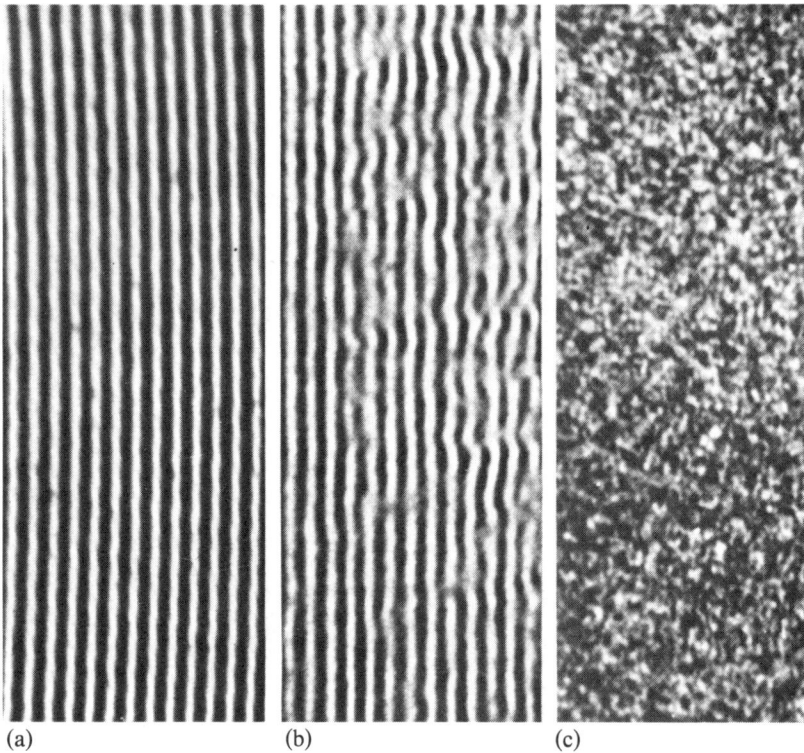

Figure 43 Microscope photographs showing various degrees of modulation of hologram fringes (a) carrier fringes only—no object present; (b) only slight modulation–simple object; (c) highly modulated fringes—complex object.

object. (Obviously the diffraction angle must not be allowed to exceed the carrier frequency beam-angle.) We can say, by analogy with radio transmission, that information about the object is diffracted into a spatial frequency *band* on either *side* of the carrier frequency. Hence the term *side-band*. Figure 43(c) shows how a highly complex, three-dimensional object can modulate the carrier fringes to such an extent that the latter fringes are not easily discernible by eye. In this case, both the amplitude and the frequency of the fringes are modulated in both the x and y directions.

4.9 The mathematics of holography

I hope that the description of holography given in the previous few Sections has given you some feel for how and why holography works. But in those Sections we only considered how *transmission* objects diffracted the laser beam and formed interference fringes with the reference beam. In fact, we needn't have been so restrictive. If we generalize the arguments by expressing them mathematically, we find that we need say nothing about the nature of the object whatsoever; the theory works just as well for reflection objects as for transmission objects.

Suppose that we represent our reference beam when it reaches the photographic plate by the expression

$$\psi_R = A_R \cos [\omega t + \varphi_R (x, y)] \qquad (15)$$

where A_R is the amplitude of the reference wave, and $\varphi_R (x, y)$ is the phase of the reference wave at all points (x, y) on the photographic plate.

In the same way, let us describe the complex wave scattered (diffracted) from the object, by the expression

$$\psi_{OB} = A_{OB} (x, y) \cos [\omega t + \varphi_{OB} (x, y)] \qquad (16)$$

when it reaches the holographic plate.

What are the differences between the object wave and the reference wave? Let me list them.

1 Whereas the amplitude of the reference wave (A_R) is the same at all points across the photographic plate (it's a plane wave), the amplitude of the object wave

varies from point-to-point across the plate (because of diffraction by the object). It is therefore a function of x and y.

2   In general, the phase of the reference beam (φ_R) will be either a constant (if the reference is normal to the plate) or proportional to x and/or y (if it arrives obliquely at the plate), whereas $\varphi_{OB}(x, y)$ will be a very complicated function of x and y because of diffraction by the object.

We must now ask what the light distribution at the photographic film will be when these two waves overlap each other.

We know that film can only respond to the time average of the *intensity* distribution, i.e. to

$$I = \langle(\psi_R + \psi_{OB})^2\rangle \tag{17}$$

where I have added the waves together *before* squaring because I know that ψ_R is coherent with ψ_{OB}.

Exercise 1

See if you can show, by putting the expressions for ψ_R and ψ_{OB} into equation 17, that the film responds to the light distribution

$$I = \frac{A_R^2}{2} + \frac{A_{OB}^2}{2} + A_R A_{OB}\left[\cos(\varphi_R - \varphi_{OB})\right] \tag{18}$$

Hint   You will need to use the trigonometrical identity

$$\cos(X + Y) + \cos(X - Y) = 2\cos X \cos Y.$$

If we expose the photographic plate in such a way that the *amplitude transmittance* of the developed plate is proportional to the *incident intensity* given by equation 18, and if we use a playback wave ψ_P whose phase (i.e. direction and curvature) is identical to that of the original reference wave, then the final wave transmitted by the hologram (which I shall label ψ_T) is given by

$$\psi_T \propto \frac{(A_R^2 + A_{OB}^2)\psi_P}{2} + \frac{A_P A_R A_{OB}}{2}\cos(\omega t + \varphi_{OB})$$

$$+ \frac{A_P A_R A_{OB}}{2}\cos(\omega t + 2\varphi_R - \varphi_{OB}) \tag{19}$$

Exercise 2

Can you show that this is the case? You will again need to use the identity

$$\cos(X + Y) + \cos(X - Y) = 2\cos X \cos Y.$$

We can now make some observations about the three terms in the transmitted wave ψ_T.

The first term is simply an attenuated form of the playback beam, and as such it will be along the axis of the playback beam. Notice though that A_R was constant across the photographic plate, whereas A_{OB} was a function of x and y. Hence the attenuation will not be uniform across the hologram, and consequently the straight-through beam will be subjected to some slight diffraction. This effect will be small if $A_R \gg A_{OB}$. It is this variation of A_{OB} with x and y which is equivalent to the 'self-interference' of diffraction orders with themselves that I was discussing in Section 4.8 (see Fig. 44).

The second term in the expression for ψ_T is identical to the object wave with which we started (equation 16) except for the multiplicative constant $\frac{A_P A_R}{2}$. The hologram and playback beam have, therefore, generated between them a wave, which is in every way identical to that which would have originated—nay, did originate, from the real object itself. We have succeeded in storing on the hologram, information which allows us to reconstruct the original optical field.

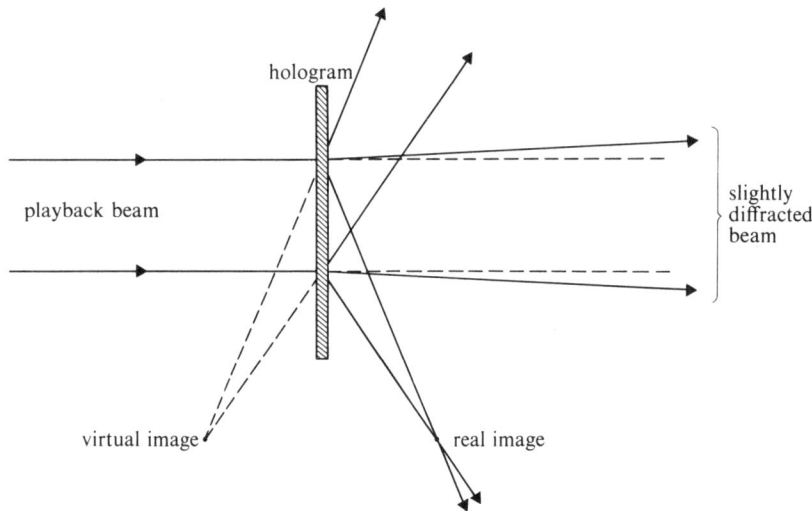

In addition, we have generated another wave (the third term in ψ_T) which looks very similar to the object wave, but which is not quite the same. Again we have the simple multiplicative factor $\left(\dfrac{A_P A_R}{2}\right)$ which is of little consequence—it merely modifies the total brightness. More important is the relevance of the difference between $A_{OB} \cos(\omega t + \varphi_{OB})$ and $A_{OB} \cos(\omega t - \varphi_{OB} + 2\varphi_R)$.

To begin with, the latter expression has an extra phase term, $2\varphi_R$.

This term will separate this output wave from the original object wave direction by an angle $2\varphi_R$, and thus from the reference (or playback) wave by an angle φ_R (Fig. 45).

But also, and perhaps more interestingly, the object phase is reversed in this output wave (i.e. φ_{OB} has become $-\varphi_{OB}$). What does the image look like when the phase term is reversed? Well, if you reverse the phase of a wavefront, you invert its shape. In fact, it is this phase reversal which is responsible for the *pseudoscopic* nature of the real image.

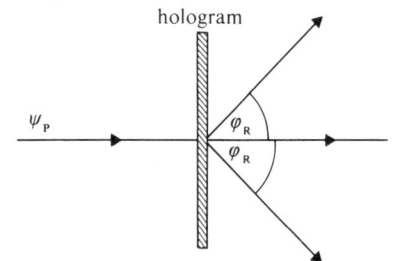

Figure 45 The angular relationship between the direction of propagation of the playback wave, the real object wave, and the virtual object wave.

5.0 Holography: Practical considerations — *How to satisfy the theoretical requirements*

5.1 What needs to be considered?

Up to this point we have considered only the principles involved in holography. But if holography is to be a practical proposition, we must make sure that the assumptions in our theoretical treatment in the preceding Section, can be realized in terms of experimental hardware. The photographic film obviously plays a central role in practical holography, and in many ways the ultimate success of any holographic set-up depends upon the suitability of the film. At least three properties of the film are likely to be critical, namely film linearity, resolution and spectral sensitivity, so we shall examine these properties in some detail. We shall then go on to look at the stability and coherence conditions needed for a holographic set-up.

5.2 Film linearity

An assumption which I made several times in preceding Sections was that the film was capable of converting the time-averaged *intensity* distribution of the optical disturbance at the detecting plate into an *amplitude transmittance* of the emulsion that is linearly related to this intensity distribution. This requirement was necessary if the playback was to reproduce the optical field accurately. But it is not an obvious consequence of simply exposing the plate. Let's see why.

You will remember from Unit 2, that film manufacturers generally provide 'characteristic curves' for their film of the kind shown in Figure 46. This plots the

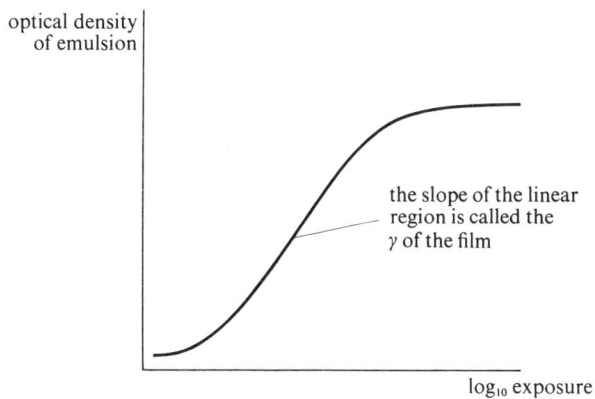

Figure 46   Film manufacturers usually specify the performance of their films in terms of the 'Hurter-Driffield' curves (sometimes called 'gamma curves').

optical density of the developed emulsion against the log of the exposure (i.e. light intensity × time), and shows both saturation of the emulsion at high exposures, and basic fog on the emulsion at zero-exposure. But this kind of plot is not very helpful for holography; a more useful curve is shown in Figure 47. Here the *amplitude transmittance* of the emulsion (which can obviously be calculated from the emulsion density) is plotted against the exposure. Since the exposure of the film is directly proportional to the light intensity at the film, the straight part of this curve is the region over which the amplitude transmission is linearly related to the optical intensity, i.e. the region over which we can satisfy the requirement for accurate holography. In particular (as can be seen from Fig. 47) it is that region for which a

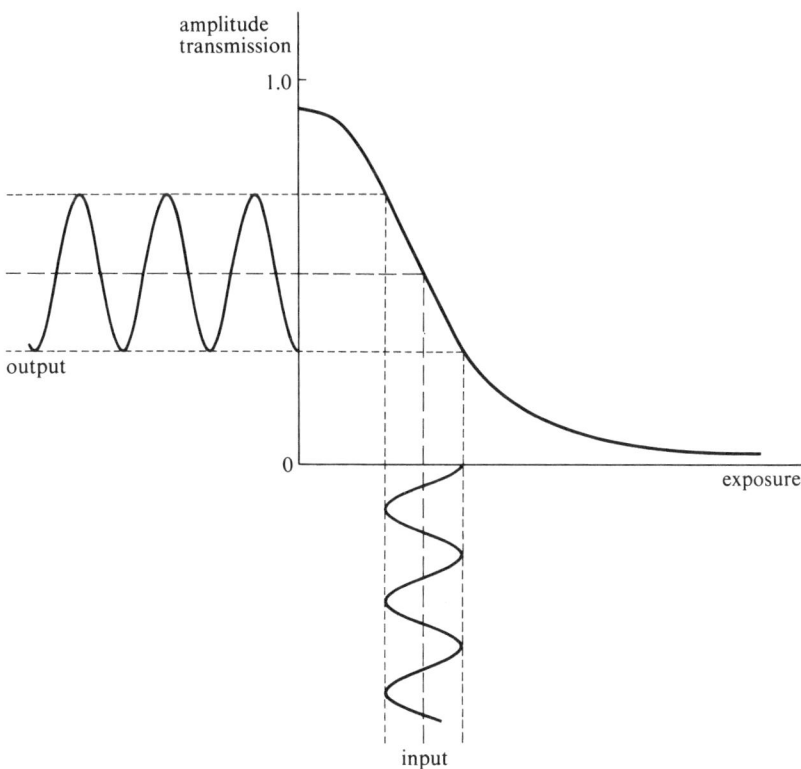

Figure 47   The amplitude transmission/ exposure curve. When all the input signal lies on the linear part of the curve, an undistorted, amplified output is obtained.

sinusoidal *intensity-distribution* of the optical field will be stored as a sinusoidal *amplitude-transmission* distribution on the film, without distortion of the sinusoidal shape. If for any reason, however, the sinusoidal input strays onto a non-linear part of the curve, the output will be distorted. In 'Fourier language' this implies that other harmonics will be introduced into the amplitude transmission, and these other harmonics will generate extra plane-wave components on playback. Since these plane waves formed no part of the original object field, they can only serve to distort the final image.

Obviously the ideal would be to record fringes which, on reconstruction, yield an image of maximum brightness and minimum distortion. Unfortunately these two requirements are mutually incompatible. Consider the consequence of striving after maximum brightness. This would be achieved by generating fringes with as high an amplitude modulation as possible, that is, by making the object-light amplitude the same as the reference-beam amplitude.* How does this work? Well, you saw in Section 4.9 that the time-averaged intensity is given by

$$I = \frac{A_R^2}{2} + \frac{A_{OB}^2}{2} + A_R A_{OB} \ [\cos(\varphi_R - \varphi_{OB})] \qquad \text{(Eq. 18)}$$

It is the last term here which generates the fringe pattern, because where the object and reference light is in phase (i.e. where $\varphi_R - \varphi_{OB} = 0$) the interference is constructive and this last term is *additive*. On the other hand, destructive interference results at positions where the object and reference light is 180 degrees out of phase (i.e. where $\varphi_R - \varphi_{OB} = 180°$). In this case

$$\cos(\varphi_R - \varphi_{OB}) = \cos 180° = -1 \qquad (20)$$

and the last term in equation 18 is *subtractive*.

Hence I varies from a maximum of

$$I_{max} = \frac{A_R^2}{2} + \frac{A_{OB}^2}{2} + A_R A_{OB} \qquad (21)$$

to a minimum of

$$I_{min} = \frac{A_R^2}{2} + \frac{A_{OB}^2}{2} - A_R A_{OB} \qquad (22)$$

When the object and reference amplitudes are equal, so that $A_R = A_{OB}$, then the depth of modulation of the fringes $(I_{max} - I_{min})/(I_{max} + I_{min})$ is a maximum. The time-averaged intensity will vary from zero to $2A_{OB}^2$. Although this would produce maximum brightness in the image, a glance at Figure 48 shows what a disastrous effect this would have on the fidelity of the image. This kind of distortion is known as 'clipping', and it will occur whenever the fringes are over-modulated. It is clear that we must never let the destructive interference between the object and reference light reduce the resultant amplitude (and hence the exposure of the film) to zero. The film will always be very non-linear in the region of zero exposure.

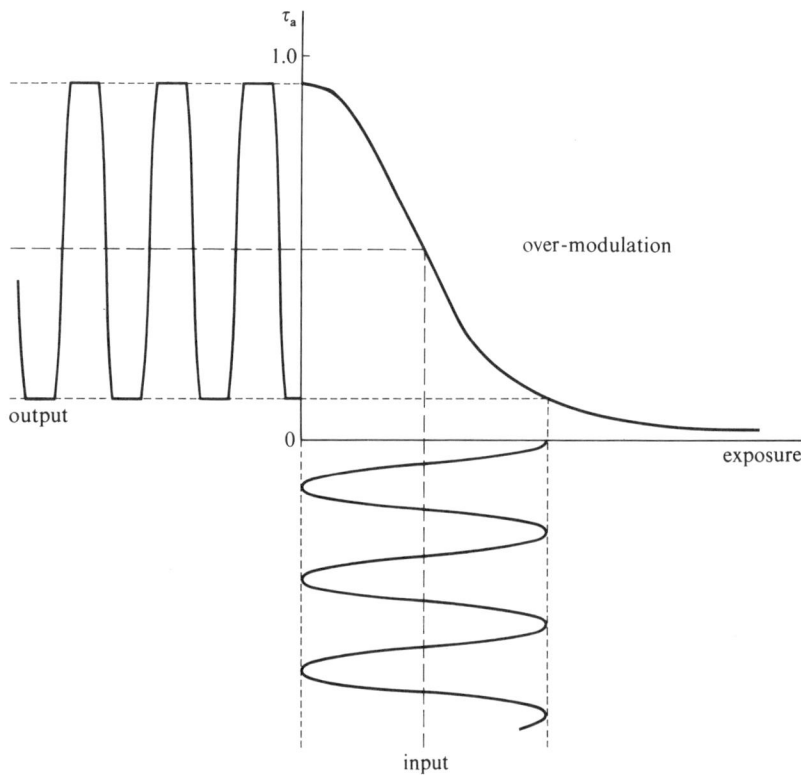

Figure 48 Although the *mean* value of the input signal lies on the linear part of the curve, over-modulation drives the output signal into 'clipping' at the extremes of its oscillation.

So the obvious thing to do is make the amplitude of the reference light larger than the amplitude of the object light.

*Remember that A_R must always be at least as big as A_{OB} in order to remove all possible phase ambiguity.

SAQ 10   What will be the extremes of I (in units of A_{OB}^2)

(i)   when the reference *intensity* is twice the object *intensity*;

(ii)   when the reference *intensity* is nine times the object *intensity*?

Calculate the *depth of modulation* in both these cases.

In practice, the optimum ratio between the object and reference light intensity will depend on the precise characteristics of the film used. Generally I_R/I_{OB} is chosen to lie somewhere between 2 and 9.

But even if we get this ratio of object to reference light correct, we can still distort the image by over-exposure. Exposure is the product of light intensity and time, so simply by exposing the film for too long we can push the input to the transmission/exposure curve into the non-linear region at high exposure (Fig. 49).

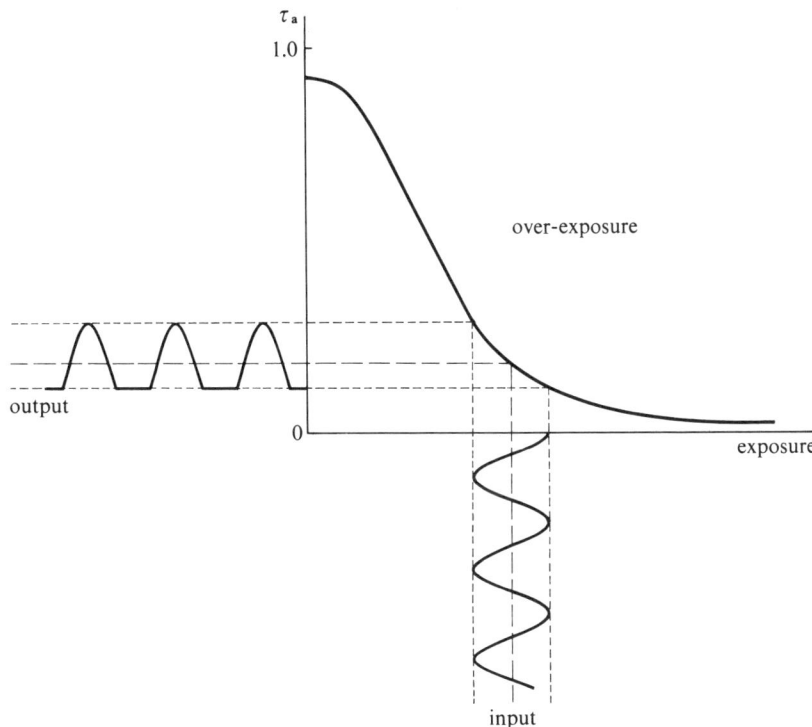

Figure 49   Over-exposure shifts the input sinewave to a nonlinear part of the curve. The output is distorted.

Notice that this also has the bad effect of negating the natural amplification properties of the hologram/film system (in Fig. 47 the output was bigger than the input), so producing a weak image of low contrast. When we make a hologram we must, it seems, pay careful attention to the exposure time as well as to getting the reference/object beam intensity ratio right.

5.3   Resolution and film speed

We can make a hologram on any material which is in some way sensitive to light. Indeed a lot of research effort is currently going into finding alternatives to photographic film—alternatives such as thermoplastic or photochromic materials.

But all these substances must have one thing in common; they must all be capable of *resolving* the fringe structure. In the early days of holography this was a major drawback. Film was capable of resolving typically of the order of two or three hundred lines per mm, whereas for an average object-reference beam angle of 45 degrees in a holographic set-up, the typical fringe (spatial) frequency is given by

fringe resolution

$$q = \frac{1}{d} = \frac{\sin 45°}{\lambda} \quad \text{(see Section 4.2)}$$

$$= \frac{0.707}{633 \times 10^{-9}} \quad \text{lines per metre*}$$

$$\approx 1\ 100 \text{ lines per mm.}$$

*I have put $\lambda = 633$ nm since this is the wavelength of the helium-neon laser light.

To attain this kind of resolution, extremely fine-grain film has to be used. The trouble with fine-grain film is that it is very slow, thus necessitating long exposure times—often up to several minutes. This is not very satisfactory when 'holographing' objects which might move! The obvious remedy is to increase the laser intensity to compensate for the low film speed. But even this apparent way out has its problems. Firstly, it might well be dangerous to increase the laser power if any of the light is going to fall on living tissue. But perhaps even more relevant is the fact that until very recently the high-power, short-duration pulsed lasers (such as ruby and neodymium) needed for *holographic 'life-studies'* were not available with long enough coherence lengths to make them suitable for this kind of work. So the photograph of the holographic image shown in Figure 50 demonstrates a considerable advance in laser technology in recent years.

holographic life-studies

Figure 50 A hologram of a roomful of people, made with a high-power, pulsed ruby laser.

5.4 Spectral sensitivity

The other problem that goes hand-in-hand with the problems of film speed and resolution, is that of *spectral sensitivity*. The vast majority of films are most sensitive to light in the green or blue part of the spectrum. Unfortunately, the commonest laser source for holography, the He-Ne laser, emits red light (at a wavelength of 633 nm). The ruby laser emits light even further towards the red-extreme of the spectrum (at 694 nm), and the neodymium laser wavelength is right out in the infra-red region, at 1 060 nm. If we are already struggling with slow film speeds, then using illumination of a wavelength to which the film is barely responsive is simply going to compound our difficulties! Fortunately, the film manufacturers have worked very hard at producing emulsions which minimize these drawbacks, so much so that the special holographic films now available have a speed, a resolution, and a sensitivity to the red part of the spectrum which might well have been pronounced quite impossible five or ten years ago.

spectral sensitivity

5.5 Stability

In Section 5.3 we saw that a typical carrier-fringe frequency, corresponding to an angle of 45 degrees between the reference and object beams, would be of the order of 1 000 lines per mm. In practical holography the angle is likely to be nearer 90 degrees, and taking this into account, together with the fact that the information is impressed upon the carrier frequency by increasing and decreasing the carrier angle, a fringe frequency of 2 000 lines per mm or more, is not unlikely. This means a fringe spacing of less than *one micrometre*—a thousandth of millimetre! This gives you some idea of the importance of the *stability* of a holographic system. A movement of any of the components—including the object itself—of more than a

stability requirements

thousandth of a millimetre could completely wash out the fringe structure, if that movement occurs during the time of the exposure. Observations like these sometimes make me wonder how holograms are ever made! Of course the obvious solutions are (i) to try to isolate the system from outside vibrations by using vibration-damping supports, and (ii) to keep the exposure time as short as possible.

For live objects, this last technique is absolutely essential (for obvious reasons). The exposure time for the hologram in Figure 50 for instance, was less than one millionth of a second. When the exposure time is as short as this, the vibration damping need only eliminate very high-frequency vibrations. (The low-frequency ones never have time to make their influence felt.)

It is also important not to forget other sources of fringe disturbance such as air movement, or differential temperature changes in any part of the light path. Although with these kinds of effect, once the source of the trouble has been recognized, the remedy is usually very simple.

5.6   Coherence requirements

By now you are probably so aware of the necessity in holography for the reference light to be coherent with the object light, that a Section entitled 'coherence requirements' might look somewhat redundant! After all, I pointed out earlier that the art of holography remained more-or-less dormant after Gabor's initial work, until the laser came along as a source of coherent light. So don't we only need to say that we *must* use coherent illumination for holography? Well, it's not quite as simple as that. You saw in Unit 5 that perfect coherence (or perfect incoherence for that matter) is an ideal which we never achieve in reality. We must talk instead of the degree of coherence of the illumination. Furthermore, we saw in the same Unit that we could quantify the degree of coherence by specifying the coherence length (or time) of the illumination (essentially a measure of the temporal coherence), and the coherence area of the illumination (which tells us something about the spatial coherence). What do we need to say about coherence length and coherence area of our illumination for holography? The answer to this question must depend to some extent on our particular system. In Figure 51, I have indicated the basic *optical*

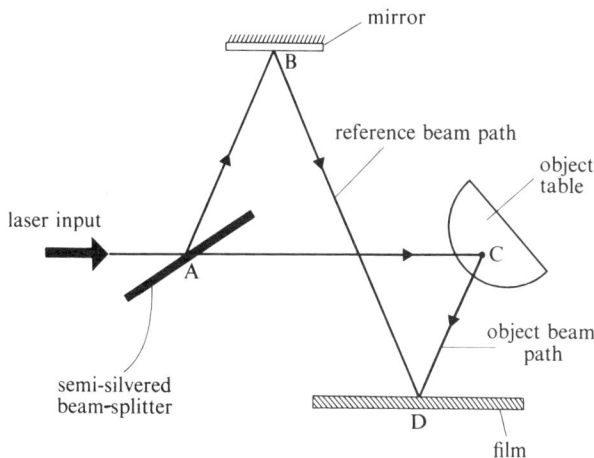

Figure 51   The coherence length of the laser must be greater than the maximum path difference between the reference and object beams.

layout that you will use when you come to make your own holograms with the bits-and-pieces from your Home Experiment Kit. It is clear that if the light in the reference beam is to interfere with the light from the object, then the difference between the reference-beam path length and object-beam path length (nominally [AB + BD] − [AC + CD] in Fig. 51), must be less than the coherence length of the laser. This means in particular that the coherence length must be *at least* as great as the depth of the object, if we want to record the whole of this depth on the hologram. The coherence length of a low-power He-Ne laser *can be* several metres, so this should pose no problems for us.* But generally speaking, as the power of the laser increases, the coherence length is reduced. Note therefore, the very real achievement in laser technology represented by Figure 50, where the depth of field is clearly many metres, and where the illuminating source was a high-powered pulsed ruby-laser.

*Your Home Kit laser is not quite as good as this. Typically, it has a coherence length of about 30 cm.

We can specify the coherence area required in a holographic system in a similarly simple way. Obviously the coherence area of the illumination must be greater than the transverse size of the object itself. Again admire the technological advance demonstrated by Figure 50, where the object is room-sized!

6.0 Holography: The technique — *How to make holograms*

6.1 Introduction

By now you should have some idea of how holography works. But more importantly, you should also have a pretty good idea of the things to look out for when you attempt to make and view holograms—which is just as well, because in this Section that is precisely what you are going to do—make and view your own holograms.

I suggest you first read through Sections 6.1–6.9, trying to understand the operations required in the experiment, and how they relate to components in the Home Kit. Then try the experiment itself, referring to the text at all the crucial points. Don't begin the experiment unless you feel you have enough time to complete it all in one go. This is particularly important with regard to the developing of your film. You may well be tempted to leave the developing until the next day. **Don't!!** The image on the film fades slowly after exposure. With ordinary film this is not particularly noticeable, but with the especially high-resolution, red-sensitive film we have provided, a delay of 24 hours could mean the total loss of your hologram. (I speak from bitter experience; this is exactly what happened to me the first time I tried the experiment!)

You will probably need a 'free run' of about three hours for a first attempt at this experiment. I suggest that you only expose one film cassette on this run. If anything goes wrong the first time, examination of your exposed film might provide some clues as to what to do to rectify the situation on a second run-through. (If you get immaculate holograms the first time, don't worry—we can't all be perfect!) Actually, I found that making my own holograms was an exciting, highly satisfying experience, and one which I heartily recommend you don't miss out on.

6.2 Preliminaries

When you take a photograph of an object or scene, you have to take care not to allow any *stray* light to fall onto the photographic film. The same sort of precautions are necessary when you make a hologram of an object. Here, the only light which should fall on the holographic plate is the reference-beam light, and the coherent light reflected from the object. What this usually means in practice is that holograms have to be made in a darkroom.

Obviously, not all of you will have access to a darkroom, so the conventional way of making large-scale holograms is not going to be available to you. Instead, we have modified the experiment slightly, so that the majority of the components are placed inside a dark box—the 'holobox'. You can see the experimental set-up in Figure 52.

When the holobox lid is in place, light can get onto the film only by passing through the relatively small hole which admits the laser light. You should find that, provided your room is reasonably dark (i.e. you don't have the room lights on and you have the curtains drawn), virtually no stray light will reach the film plane at all. You will even be able to use a small hand-torch during the experiment, without fogging the film.

The other problem which usually prevents the 'amateur' experimenter from making his or her own holograms is the difficulty of finding rigid, vibration-proof working surfaces on which to mount the optical components. (Recall the arguments I advanced in Section 5.5.) The 'professionals' tend to use heavy concrete blocks, supported by thick rubber bricks or mats. I suspect that these components are not likely to be readily found in the average family home! (Nor can they be bought for just a few pence!) Fortunately, several factors in the construction of the holobox and the layout of the experiment seem to have conspired to 'let us get away with it'

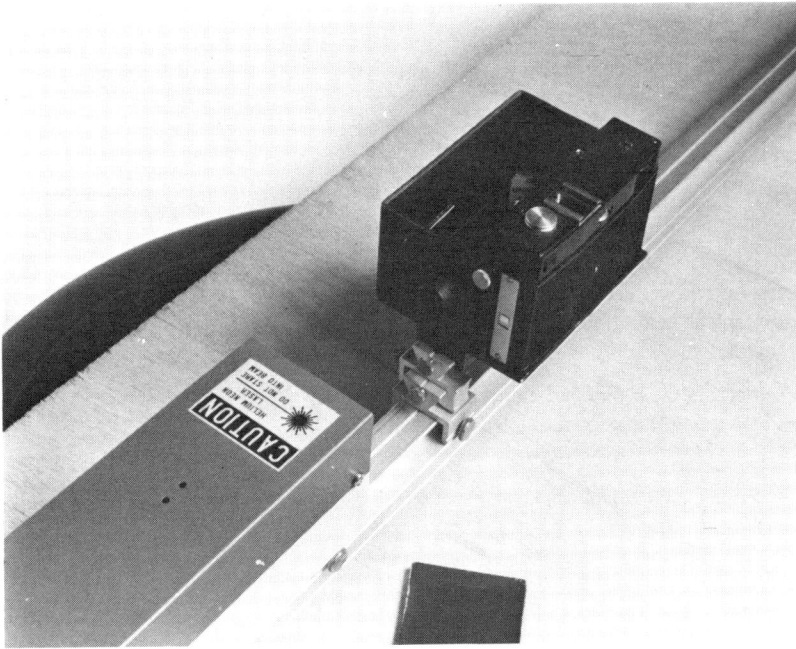

Figure 52   The optical bench set-up for making your own holograms.

with regard to these vibration problems. The miniaturization of the whole operation is obviously one of the factors, as is the fact that all the components (with the exception of the laser) are mounted in the same vibrationally dead (comparatively speaking) die-cast box. In addition, both the laser and the holobox are mounted on the same optical bench, which of course has good vibration-damping rubber feet. Anyway, whatever the reason, I found concrete blocks unnecessary—and I refuse to tempt the wrath of the Gods by inquiring too closely into the matter!

However, although you would very probably obtain perfectly satisfactory holograms with no additional vibration-isolation, it might be prudent to add just one extra layer of protection, if at all possible. You have only two cassettes of the special holographic film, and that doesn't give you much leeway for error. I found that I could get extra vibration-protection by putting the optical bench on a four-foot plank of chipboard, and supporting the chipboard on an old, semi-inflated car inner-tube. (Ask your local garage about this last item. My garage was perfectly happy to let me have an old, patched-up tube for only 25p; they even pumped it up to the right pressure for me, at no extra cost!) Figure 53 should give you some idea of the system I used. You may be able to think of alternative ways of supporting the optical bench. Perhaps an inflatable water raft, or similar item would do. Use your ingenuity.

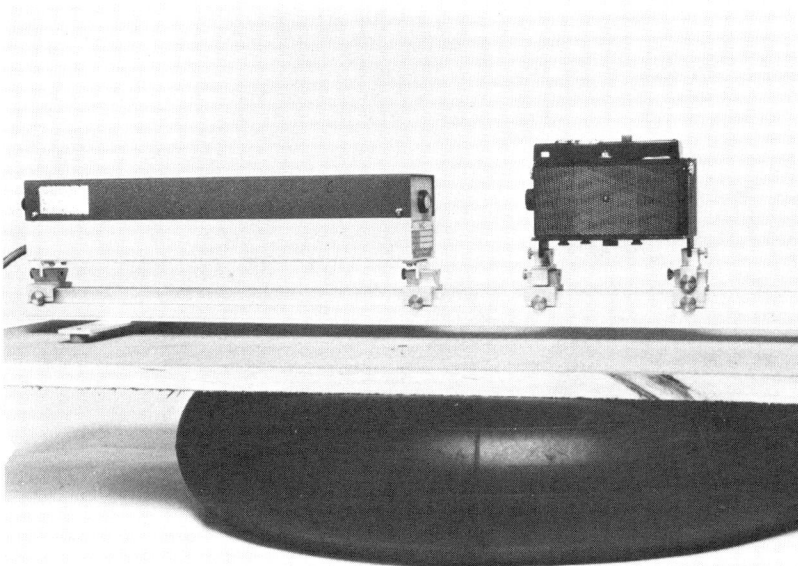

Figure 53   A car inner-tube tyre, and a plank of chipboard can be used to provide some degree of vibration isolation.

Finally, do use your discretion in choosing the best time of day to try this experiment. In general, late evening is likely to be best (from both an extraneous light, and excessive vibration point of view). But avoid the obvious problem times. If you live in a city, wait for the traffic vibrations to die down a bit—don't attempt the experiment during the rush hour! (And don't try making holograms if the Electricity Board are digging up your front drive with pneumatic drills!)

6.3 Preparation of the 'holobox'

I now intend to describe the detailed setting-up procedure for the equipment. In this section I want to describe briefly the 'mechanical' setting-up procedures; in the next Section, I shall describe the optical alignment.

The film transport mechanism for the holobox is the Home Kit camera. But before the camera can be attached to the box, you must first remove the lens and shutter mechanism.

This is a relatively easy operation. If you open the camera-back, you will see two very small screws—one above, and one below the lens (Fig. 54). These screws should be removed (using the small screwdriver in the Home Kit), and the whole of the lens/shutter mechanism separated from the rest of the camera body. *Do not lose the two small screws*. Wrap them in Sellotape and put them, along with the other screws, in a compartment in one of the moulded plastic trays in the Home Kit.

Figure 54 When these two screws are undone, the lens and shutter can be separated from the camera body.

The camera body can now be attached to the holobox as shown in Figure 55. The big screw enables you to lock the camera to the holobox from below; the two smaller screws shown in the Figure should be made finger-tight so as to hold the camera firmly in place. (Do not load film into the camera just yet.) The holobox can now be mounted on the optical bench, with its light-input hole facing the laser. In order to get the height of the holobox to match the height of the laser beam, you will have to use four saddles in all—two stacked one on top of the other, at both the front and back of the box. Position the holobox about 10 cm away from the laser.

Figure 55 The camera body is held in position on the 'holobox' with three screws.

Figure 56 Arrangement of components inside the holobox.

Refer now to Figure 56. The beam-splitter and front-silvered mirror should be mounted in their supports as shown. Never touch the optical surfaces with your fingers. *Always hold these components by their edges.* The miniature chesspieces should be placed in the holes in the object-table in such a way as to make a good display when viewed through the camera-back. You might like to spray one or two of the chesspieces with silver paint (before fitting into the object-table, of course!). The metallic sheen shows up quite differently from the ordinary white plastic when the hologram is replayed. You will need to purchase a small aerosal spray-can of silver paint if you are going to do this. Alternatively, you might like to use something metallic as one of the objects—perhaps one of the small screws from a spare optical saddle or pillar. You can mount it on the object-table with a bit of Blu-Tack. It is also useful to stick a bit of white sticky paper to the back-plate of the object-table. You can then put some lettering on this (black lettering for best results) to act as part of the object (see Fig. 57).* The lettering will help to emphasise the parallax effect when viewing the reconstructed holographic image, since different groups of letters will be obscured by the chesspieces as your head takes up different viewing positions.

(a)

(b)

(c)

Finally, do feel free to substitute any alternative objects of your own. The only requirements are that they should be small enough to fit on the object-table, and light-coloured enough to reflect the red laser-light. (Spray with silver paint if necessary!) Historically, experimenters in the field of holography have been obsessed with a desire to litter their holograms with images of chesspieces—a historical precedent to which I thought it wise to defer throughout these Units! But *you* need feel no such constraint. Try a little toy car, or a model soldier, a petit earring, or a small tie-pin instead (see Fig. 57).

Figure 57 Some suggestions for possible holographic subjects. (a) miniature chesspieces; (b) a model car (painted silver); (c) a tie-pin stud and shirt button.

6.4 Optical alignment

You should carry out your initial optical alignment with the *un*diverged laser beam. First, adjust the transverse position of the holobox so that the laser beam enters the input hole centrally, strikes the beam-splitter centrally, and passes over the approximate centre of the object-table. Adjust the height and/or the angle of either the laser or the holobox until the laser beam traverses the box at a constant height above its base. (As a rough guide, it should enter the input hole centrally and strike the back-plate of the object-table about half way up.) Now adjust the angle of the beam-splitter so that the *reflected* beam strikes the front-silvered mirror centrally.

*Incidentally, if you find any of these things done already, it simply means somebody has been there before you!

(Hold a bit of tissue paper in front of the mirror to see where the spot is.) Avoid touching the optical surfaces with your fingers. When the spot strikes the centre of the mirror, the angle of the mirror should be adjusted until the laser beam is reflected back towards the centre of the film. If necessary, open the camera-back and stick some greaseproof paper across the square hole in the camera body. This hole will form the frame for the film, and the greaseproof paper will be approximately in the plane of the film. The front-silvered mirror in the holobox should be adjusted until the laser spot is centred in this film frame. If you are in any doubt about this basic alignment, refer back to Figure 51 (in Section 5.5) where these laser-beam paths are shown schematically.

Next, you must expand the laser beam so that it (a) illuminates the whole of the object; and (b) provides a reference beam which covers the whole of the film frame.

Attach the diverging lens A to the front of the laser casing where the beam emerges. The laser beam is now diverged into a cone of light. Adjust the position of lens A on the end of the laser, and also the distance between the laser and the holobox, until the light strikes all the optical components centrally, and just illuminates the whole of the object. Check the light falling on the greaseproof paper in the film plane. The illumination should look roughly uniform across the whole plane. Make further mirror or beam-splitter adjustments as necessary.

You will almost certainly see some dark swirls across the film plane. These are caused by imperfections in the small diverging lens. Try rotating this lens about its axis to minimize the effect of such swirls. If they appear to be very bad, the lens, or even the output window of the laser, might be a bit dirty. **Clean with care**—using a 'cotton bud' soaked (if available) in isopropyl alcohol (or tape-head cleaning fluid). Dry with a clean 'cotton bud'.

6.5 Getting the reference/object beam ratio right

When you are satisfied with the optical alignment, you can make a start on getting the correct intensity ratio between the object light and the reference beam. Remove the greaseproof paper from the film plane in the camera, and position the light-sensitive diode (the photometer head) in this same plane at the centre of the 35 mm frame (Fig. 58). Block-off the laser input to the holobox with some thick black card. Only ambient room light (which should not amount to much as the room should be in at least semidarkness) and perhaps a little bit of scattered laser light will now be falling on the light-sensitive diode. Adjust the photometer-zero under these conditions. (Do this on a sensitive range, say 0.03 μA). Now select the 100 μA

Figure 58 The photometer head should be positioned in the film plane.

sensitivity range, and remove the card blocking the laser light. Increase the sensitivity of the photometer (i.e. reduce the current required for full-scale deflection) by turning the range-switch anti-clockwise until you get a sizeable (say about half-scale) reading on the meter. Record this reading. Check the zero again by blocking off the laser input. Now block-off the reference beam light reaching the film plane, by inserting a piece of black card between the beam-splitter and front-silvered mirror. (White card is no good, because a considerable amount of laser light will be diffusely reflected from it onto the photodiode.) Take the new photometer reading, now that only object-light is falling on the diode. This reading could well be as much as twenty-times smaller than the previous one. Since the photometer reading is proportional to the intensity of the light falling on the photodiode, the ratio of these two readings that you have just taken, is the ratio of the intensity of reference *plus* object light, to the intensity of object light alone.

Obviously, if we are to aim at a nominal reference to object intensity ratio of about 5:1, we must attenuate the reference beam considerably. In your Home Kit you are provided with a neutral-density filter with a single-pass attenuation of about two. If this is inserted between the beam-splitter and the mirror,* then the light will pass through the filter twice. Hence it will attenuate the reference beam by about a factor of four.

What new reference/object intensity ratio does this give you?

Your ratio should now be somewhere in the region of 5:1. But don't just trust your calculation—*measure* the beam intensities again with the neutral-density filter in place. You can alter the object-light intensity over a limited range by rotating the object-table slightly, but if the ratio of the two beam intensities lies somewhere between about three and ten, then it's probably best to leave well alone—you'll have no trouble in getting some kind of hologram.

When you're happy that the intensity ratio at the centre of the film frame is all right, it's worth just checking that other parts of the film frame area are receiving light in the correct proportions also. This is where object-table rotation can really serve a useful function—the uniformity of object light across the film can be very dependent on the particular angle the object makes with the film plane, especially if there are mirror-like reflecting surfaces on the object. When you think you have adjusted everything as well as you can, have a last look through the camera-back at the illuminated object. Move your head about a bit. What you see now is what you ought to see when you replay your finished hologram. Put the lid on the holobox gently, without disturbing any of the components inside.

6.6 Loading the film

You are now almost ready to make your holograms. Take one of the cassettes of holographic film from the Home Kit, and draw out about 15 cm (6 inches) of the film from the cassette mouth. Take the scissors and cut along the length of the film for about ten cm, as shown in Figure 59. (This will then act as leader for the film.

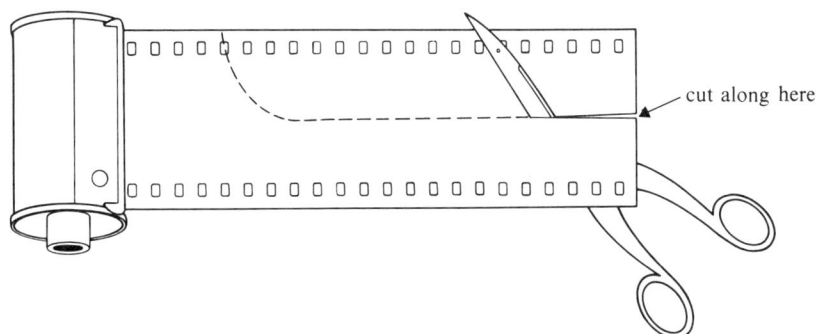

cut along here

Figure 59 Cut the holographic film so as to provide several centimetres of 'leader'.

*Stick it to the base of the holobox with Blu-Tack, just in front of the mirror, and at such an angle that any light reflected from the front face of the filter will not fall onto the film.

When rewinding the film into the cassettee, there will be less likelihood of your winding too far—the frame counter will stop rotating backwards as soon as this leader part of the film is reached.) Now load the cassette into the camera in the usual way. Close the camera back.

Block-off the laser beam with black card placed in front of the holobox input hole. Wind the film on a frame. Push the shutter button on the top of the camera, and wind-on a second frame. Remember that one frame of the film is exposed to the inside of the holobox all the time. The film speed is very slow however, so you should have little trouble in a darkened room.

6.7 Exposure

You are now ready to expose the film to the laser light. As you have ten frames available, a sequence of exposure times is desirable. I suggest that you follow the sequence below. (Since the beginning of the film may be spoilt during development, it is a wise precaution to repeat the 0.5-, 1-, 2-second sequence, as indicated.)

Frame number	1	2	3	4	5	6	7	8	9	10
Exposure time (seconds)	0.5	1	2	0.5	1	2	4	8	16	32

The exposures are made simply by removing the black card, which is blocking off the laser beam, for the desired amount of time. Before you remove the card, make sure that you are not resting it on the bench or holobox—you must avoid contact with the optical bench if you are not to induce vibrations. Do not talk loudly during the exposure either—the moving air could disturb the fringes.

I found that the best way of timing the short exposures was to count to myself: one-and-two-and-three-and- etc. while watching the second-hand of a watch. Once you've got the timing into your head, an exposure of a half-second is obtained by simply removing the card from the beam on the count of 'and' and replacing it on the following number count. One, two and four seconds are easily obtained in the same way. In fact, since you aren't too worried about *exact* exposure times here, this technique is probably good enough for all your exposures.

6.8 Finishing-off

When you reach the end of the film, turn off the laser and rewind the film back into the cassette. Before you can rewind, you will have to press the trigger release button to free the mechanism. *Do not be tempted to remove the camera from the holobox before rewinding, and do not be tempted to put the room lights on.* Remember the camera-front is still open—you want as little light as possible around during rewind.

When the frame counter stops rotating, and you hear the film come off the spool, open the camera-back and remove the cassette. *Now you can put on the lights.* You must develop your film straight away—do *not* leave it overnight. Develop the film in the usual way. You should have pretty dirty developer by the time you have finished. After washing, hang the film up to dry. Leave viewing until tomorrow.

6.9 Check-list

If you took my advice, and read the previous Sections through before starting the experiment, you will now be ready and raring to go. To help you, I have reproduced a check-list of activities to be carried out during the experiment. Use it to make sure you don't forget anything. Good luck!

Check-list

1 Obtain vibration-damping equipment, and a small pen-torch.

2 Prepare any 'objects' you are going to use. Possibly spray some with silver paint. Affix lettered backdrop to the object-table.

3   Remove lens from camera.

4   Attach camera body to holobox.

5   Set up the laser and holobox on the optical bench. Place the bench on the vibration-isolation equipment.

6   Turn on the laser. (The laser takes about ten minutes to stabilize. It is best to turn it on now, and leave it on.)

7   Put the beam-splitter, front-silvered mirror, and object-table in place in the holobox.

8   Adjust the position and angle of the laser and holobox until you are satisfied that the laser beam is striking the optical components centrally.

9   Adjust the rotation-angle of the beam-splitter so that the reflected beam strikes the mirror centrally.

10   Adjust the mirror angle so that it reflects the beam back to the middle of the film frame.

11   Diverge the laser beam with lens A. Adjust the position of this lens, and the distance between the laser and holobox until the object is completely illuminated, and the reference beam fills the film frame.

12   Measure the reference/object beam intensity ratio with the photometer.

13   Insert the neutral-density filter. Measure the new intensity ratio.

14   Check the intensity ratio across the film frame.

15   Cut the 'leader' in the holographic film.

16   Load the cassette into the camera-back.

17   Block-off the laser beam, and wind-on a couple of frames.

18   Make several different time exposures—using all ten frames on the cassetted film. (If you find more than ten frames, make use of them.)

19   Turn off the laser, and rewind the film into the cassette.

20   Develop, and hang the film up to dry.

6.10   Viewing your holograms

View your holograms systematically. Don't be impatient—it might take you a while to get the viewing conditions just right. First, remove the holobox from the optical bench (you can dispense with the vibration-isolation bits-and-pieces at this stage too) and replace it with the metal 35 mm filmholder provided in the Home Kit. Slide your developed hologram film into the holder. (Incidentally your hologram film, although clearly divided up into individual frames, will not appear to contain any recognizable information. Each frame will seem to consist of random swirls and blotches and perhaps a few fringes.) Put the film in the holder with the same orientation that it had in the camera. That is, it should 'curl' *away* from you (as you look towards the laser), and the 'sticky-tape' end should be on the left-hand side, and the 'leader' end on the right-hand side. Now, place the beam-diverging lens A onto the end of the laser housing, and push the filmholder towards the laser until the cone of light just fills one complete 35 mm frame. Recall the approximate angle at which the reference beam struck the film in the camera when you were making the holograms, and twist the filmholder now so that the playback beam strikes the hologram at about the same angle. This means that you will have to rotate the filmholder about 30 degrees clockwise (as seen from above). Now look *through* the hologram, towards the right-hand side (where the objects were when you looked through the back of the camera) and see if you can see the objects. The image might be quite dim, so have the room lights off while viewing. Examine all the different exposures, and choose the frame which seems to give the clearest image. Now look for the various depth effects described in the earlier parts of these Units. As you move your head from side-to-side, do you see parallax effects? For instance, does one of the chesspieces move behind another, or are different letters in the back-drop obscured as you move about? Do the chesspieces throw shadows onto the white back-drop? Try moving your head up-and-down, as well as from side-to-side. Experiment with these images as much as you like.

You should also see if you can locate the real, projected image generated by your best hologram. When you have located the virtual image, remove the diverging lens so that the *un*diverged laser beam strikes the hologram. The real image should now be on the right-hand side (looking towards the laser) and as far in front of the hologram as the virtual image was behind it (Fig. 60). **Do not look into the hologram in this case—the laser beam is undiverged.** Instead, project the image onto a piece of white card. You will have to position this card quite accurately, because the range of distances over which the image is in focus is quite small. Try moving the film slowly through the filmholder, so that the laser beam 'samples' different parts of the hologram. The image on the screen should change.

> SAQ 11   You may have noticed that all lettering is mirror-reversed when the real image is viewed in this way. Why? How would you view the real image so as to see the lettering the correct way round?

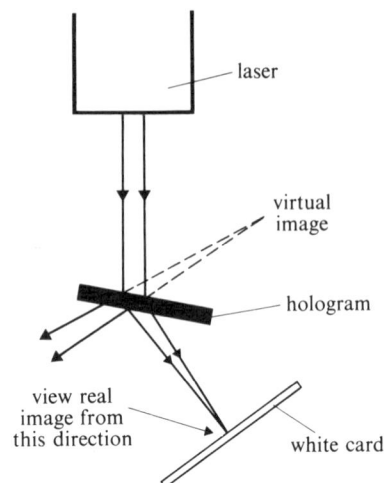

Figure 60   Locate the real image on a piece of white card.

6.11   Another try?

If you didn't succed in making very good holograms the first time, perhaps you would like to try again with the second cassette of holographic film. Now that you've been through the experiment once, you should find that you are much happier about what you should be doing and why, and you can perhaps spend a bit more time optimizing the alignment, and object orientation.

If your holograms worked out fine anyway, then perhaps you would like to try again with different objects. Remember though, that if you open the lid of the holobox while your film is loaded in the camera (so that you can change object mid-experiment for instance), then you will have to forfeit the film frame which is being exposed at that time. Make sure that you wound-on from your previous holographic exposure.

One other object which you might like to try on your last two or three frames, is the 'plane-mirror object'. You can use the glass window from one of the plastic slides as the 'mirror object'. Remove all the objects from the object-table, and position the glass from the slide up against the object-table back. Use Blu-Tack to hold the glass to the table. Now adjust the position of this table so that the object-light is specularly reflected onto the film plane. The reference beam and the object beam should now interfere to produce an approximately sinusoidal light distribution across the film plane. On development, the hologram should be a sinusoidal diffraction-grating.

> SAQ 12   How would you check that the hologram was a sinusoidal grating?

7.0   Holography: The professional touch — *Déjà vu*

7.1   A closer look at the virtual image

Before you finally put your Home Experiment Kit away, I would like you to go back to the very opening of these two Units—to the Prelude—and re-read the setting-up procedure for viewing the professional hologram described there. Having made your own holograms, and, hopefully having learnt something about the way in which holograms work, you should now be able to look at this professional hologram with a much more critical eye. Look at the virtual image again. Check the parallax effect. Convince yourself that the stopwatch and the magnifying glass really are in different focal planes. You are sure that the glass is *in front* of the watch, aren't you? The depth of field in a holographic image is very obvious when a wide-aperture photograph is taken of the various objects in this reconstructed image. Figure 61 shows this effect for a different hologram (the old chesspieces again!) in which there are three quite distinct object planes. Yet with the small depth of field available on a wide-aperture photograph, only one of these planes is in focus at any one time.

You should also try reducing the viewing area of the hologram by covering parts of it with paper or card. The whole image is still present, isn't it? But the 'window' through which you are looking has been reduced in size, so that the number of different viewpoints that you can adopt has been severely restricted. The effect is not unlike looking through a keyhole.

Figure 61   The chesspieces in this hologram are positioned in three distinct planes. When a camera is focused on any one of these planes, the other two planes are out of focus. The holographic image has real three-dimensional depth.

7.2   Image magnification

In the analysis earlier in these two Units, reconstruction of the hologram was always accomplished with *plane-wave* illumination. When we use spherical-wave illumination (as here), then we find that the size of the virtual image depends on the radius of curvature of the playback wave at the plane of the hologram.

For instance, if you push the hologram towards the laser, without making any other alterations, then the image gets smaller. Try it. As you pull the hologram away from the laser, the image gets larger. The waves coming from the diverging lens are roughly spherical, so that the radius of curvature of a wavefront close to the lens is smaller than the curvature of the wavefront a greater distance from the lens. The reconstructed holographic image is the same size as the original object, only when the radius of curvature of the reconstructing beam at the hologram is made identical to the radius of curvature which the reference beam had when the hologram was made.

Incidentally, you should be able to see exactly this same effect with your own home-made holograms.

7.3   The real image

Now examine the real, projected image. Move the hologram much closer to the laser and set up the ground-glass screen on the optical bench, about 20 cm in front of, and slightly to the left of the hologram. Remove the diverging lens from the laser. Adjust the position of the glass screen until the projected image is in focus. View the screen from the side furthest *away* from the laser. Now move the hologram about in the path of the laser beam, so that the beam 'samples' different parts of the hologram. Notice that when the beam samples the top of the hologram, the image is projected as if seen from above. When the bottom of the hologram is sampled, the viewpoint is from below. The left of the hologram gives a left viewpoint, and the right, a right viewpoint. How sharp is the focus in the real image? This is a measure of the depth of field available in this particular reconstruction process.

I would now like you to try a slightly different way of reconstructing the real image. Replace the diverging lens A, and follow it with the converging lens F. Adjust the position of lens F on the optical bench so that it produces a parallel (or slightly *con*verging) beam of laser light about 2.5 cm in diameter. Use this parallel beam to reconstruct the real image as before. You should still find that different sampling positions on the hologram give different viewpoints in the image, but how sharp is the focus in the image this time?

> **SAQ 13**   You should find that the depth of field in the reconstructed image is much smaller than it was before; that is, less of the depth of the object is in focus at any one position. Why do you think this is?

You should now find that you can easily position the viewing screen in a plane where the stopwatch is in focus, but where the magnifying glass is blurred. Move the ground-glass screen so as to bring the magnifying glass into focus.

Remember, the magnifying glass is *in front of* the watch. So which way should you move the screen to bring the glass into focus? Try it.

You have to move the screen *further away from you* to bring the magnifying glass into focus!

SAQ 14 Explain this.

7.4 Professional recording techniques

The recording layout that you used to make your own holograms was designed for ease, simplicity and compactness. The professional holographer however, is not bound by such requirements. He can afford to expand his field of view; he can use green or blue laser light (from an argon laser), and film which is much more sensitive to radiation in the green/blue parts of the spectrum; he can use much more powerful lasers—up to several watts continuous power nowadays—to keep his exposure times low; he can use a pin-hole spatial filter at the focal point of his convex-lens beam-expander (see Fig. 62) and so reduce the 'optical noise' (caused by imperfections in, and dust on, the lens) in his coherent illumination.

Figure 63 shows a typical holographic recording set-up. It sometimes comes as a surprise to realize that the light illuminating the object need not come from only one angle. The only requirement is that *all* the object illumination must be coherent with the reference light. The chessboard (sorry!) in the Figure is illuminated from both left and right.

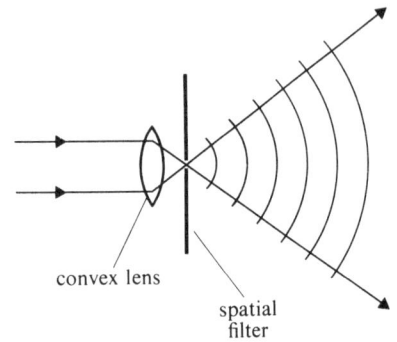

Figure 62 When a convex lens is used as a beam diverging element, a spatial filter can be incorporated in the focal plane.

Figure 63 A professional holographic-photography set-up. Note that the object is illuminated from two different directions (the light is reflected from the two mirrors at the bottom-left and right of the photograph). The reference beam falls directly onto the holographic plate at the bottom centre of the photo.

The other trick which the professionals often use nowadays, is that of *hologram bleaching*. One of the difficulties with the amplitude-type of holograms that you have made, is that the silver grains strongly absorb the playback light. If, however, the amplitude variations can be converted into phase variations, then we have a *phase hologram*. The diffraction efficiency of such a hologram is much higher than its amplitude counterpart.

hologram bleaching

phase hologram

There are many formulae available for making up suitable bleaching agents for holographic film. Unfortunately, most of these bleaching solutions use rather toxic chemicals and are really only suitable for use under proper laboratory conditions. However, current research is yielding new developing and bleaching techniques which look like making the holographic art a great deal simpler. We can almost certainly look forward in the future to even better holograms than the professional one in your Home Kit.

8.0   Holography: Colour images — *The ultimate in realism?*

8.1   The cross-talk problem

The realism of the reconstructed image produced by the hologram in your Home Kit is perhaps most impaired by the lack of colour. Is there any way in which we can record a hologram in *full colour?* At first sight this would appear to be a relatively simple problem to solve. After all, if we illuminate an object with red light, then only those parts of the object which have a high reflectivity in the red part of the spectrum will contribute to the generation of the object beam. This red object-beam will then interfere with the red reference-beam at the holographic plate and generate what we might loosely describe as 'red' intereference-fringes.* If we were now to illuminate the object with blue light, we might expect different parts of the object to have a high reflectivity in the blue part of the spectrum. Consequently the blue object-beam would be slightly different from the previous red beam, and so would lead to slightly different fringes at the hologram—'blue' fringes. Similarly, illumination of the object with green light might be expected to generate a third, 'green' set of fringes on the hologram. There is no real reason why these different-colour exposures have to be made in sequence. In practice, with suitable optics, we can illuminate the object with all three colours at once, and record the three sets of holographic fringes in a single exposure. For instance, the system shown in Figure 64, which for clarity shows the combination of only two colours, could obviously be extended to three colours without any difficulty. If we choose the three colours carefully (i.e. if we choose the three primary colours) then the object should appear as though illuminated by white light.

colour holography

*The interference fringes are not really red of course—they are black-and-white. I use these 'colour' terms to differentiate one black-and-white set of fringes from another.

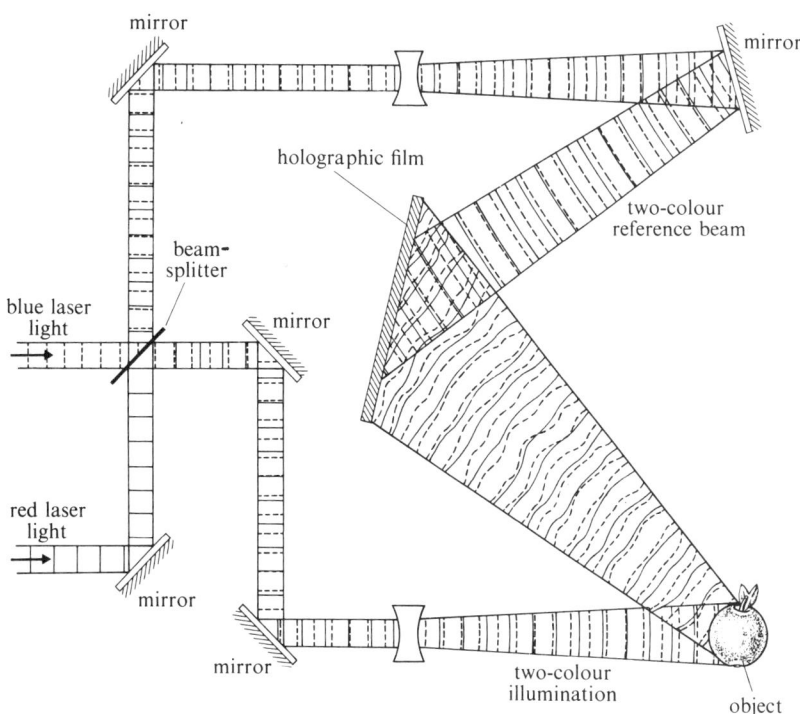

Figure 64   Making a 'two-colour' hologram.

To reconstruct the full-colour image we have only to illuminate the hologram with the three differently-coloured reference beams. But, herein lies the snag. As you saw from the television programme, a hologram made using red light can be played back quite well using green or blue light. The fringes on the hologram have no way of discriminating against the 'wrong' colour. Consequently the 'red' fringes on the hologram in Figure 64 will not only regenerate a red image when illuminated with red light, they will also regenerate a spurious blue image when illuminated with blue light. In addition, the 'blue' fringes will produce the correct blue image from the blue component in the playback beam, but will also produce a spurious red image from the red component. Of course, the red image from the 'red' fringes and the blue image from the 'blue' fringes should be superposed in the reconstruction, yielding a coloured image. But on each side of this image will be spurious, *cross-talk** images (Fig. 65) caused by the red light playing back the 'blue' fringes, and the blue light playing back the 'red' fringes.

cross-talk

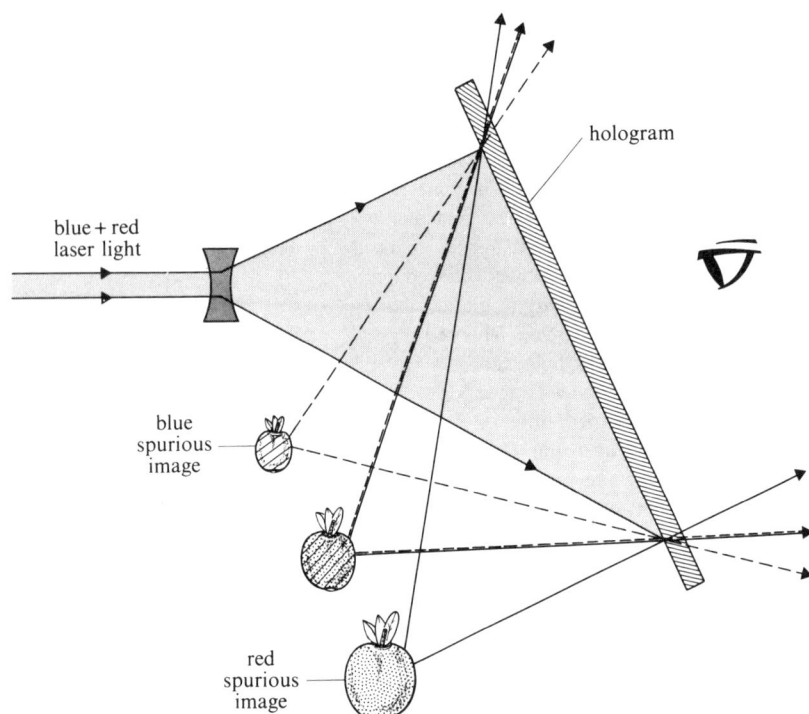

blue + red laser light

hologram

blue spurious image

red spurious image

Figure 65 Playback of a 'two-colour' hologram generates two spurious images in addition to the two-colour image.

SAQ 15 The spurious images shown in Figure 65 tend to be spatially separated from the full-colour image. Can you explain why?

SAQ 16 If a hologram is made using *three-colour* illumination, how many spurious images will there be?

8.2   Volume holograms

Surprisingly, it turns out that the only modification necessary to suppress these spurious images, is the very minor one of increasing the *thickness* of the emulsion on the holographic plate. So far in these two Units, I have assumed that the interference fringes recorded on the photographic emulsion are essentially two-dimensional—i.e. that they have no depth. But this is obviously nonsense—we cannot make 'depth-less' emulsions! A typical emulsion might be about ten microns thick, and since the fringes could well have an average spatial period of about one micron, the emulsion would be several times thicker than this typical fringe spacing. If we deliberately increase the thickness of our emulsion, then it can be as much as several tens-of-times thicker than the average fringe spacing. Consequently, the fringes will now be recorded throughout the thickness of the emulsion, as shown in Figure 66. The hologram will be three-dimensional. Such a hologram is referred to as a *volume hologram*. As you can see from Figure 66, the three-dimensional fringes in the emulsion will be oriented so that interference maxima lie along the bisector of

volume hologram

*This is another jargon term borrowed from the field of radio communications.

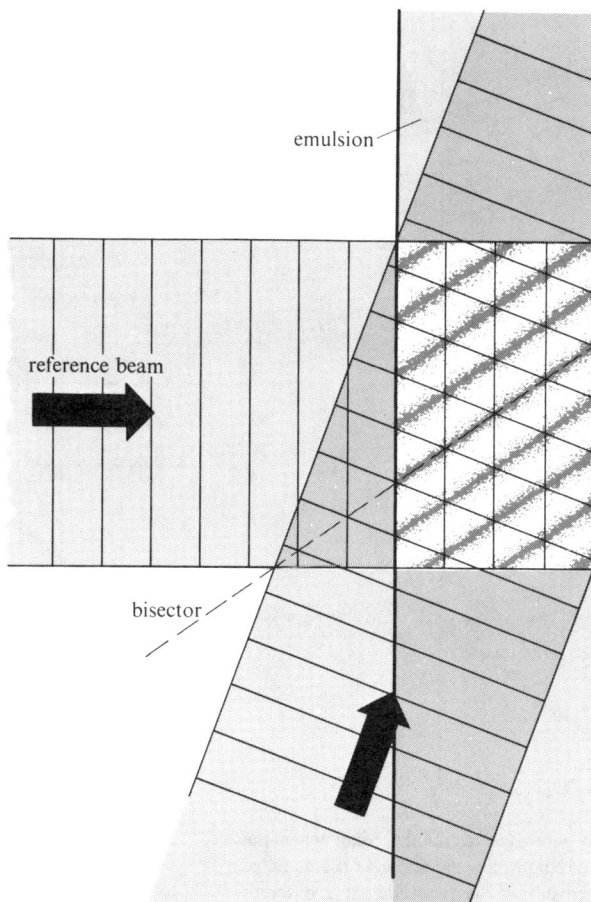

Figure 66   The fringes in a thick-emulsion hologram lie along the bisector of the reference and the object beams.

the angle between the two interfering beams. In the case shown in Figure 66, where the object beam is simply a plane wave, the resulting hologram is really a *three-dimensional* sinusoidal diffraction-grating.

In what way would you expect such a diffraction grating to deflect an incoming playback beam?

We have already met this problem, albeit in a slightly different guise, in one of the earlier TV programmes, when we showed you how the diffraction of X-rays from a three-dimensional molecule or crystal could be modelled by the diffraction of laser light in an 'optical diffractometer'. Diffraction of light by a volume hologram is very similar to diffraction of X-rays by a crystal.

In the case of the volume hologram shown in Figure 67, diffraction takes place not only at the first plane (AB), but at all subsequent planes throughout the depth of the hologram. If the light diffracted in a given direction by any one of these planes is *in phase* with the light diffracted in the same direction by all the other planes, then *constructive* interference will result, and a strong diffracted-image will be formed. As you can see from Figure 67, this is precisely what happens when the reconstructing wave has the same wavelength, and travels in exactly the same direction as the original reference beam. The wave diffracted by the plane AB is travelling in the same direction, and *in phase* with the wave diffracted by the planes CD or EF, or *any of the intermediate planes*. The diffracted wave satisfies the boundary conditions (as determined by the fringe structure on the grating) not only in the two-dimensional plane (AB) of the grating, but also throughout the depth of the grating. You should now convince yourself that this 'depth' condition cannot hold if either (i) the *incident angle* of the playback beam is changed or (ii) the *wavelength* of the playback beam is altered. In both these situations, the transmission maxima (or minima) of the grating will be in the wrong places for the contributions to the diffracted wave, from the different planes throughout the depth of the grating, to be in phase with each other. In general, this will lead to destructive interference and the diffracted wave will be suppressed. Figure 68 shows the situation when the wrong wavelength is used.

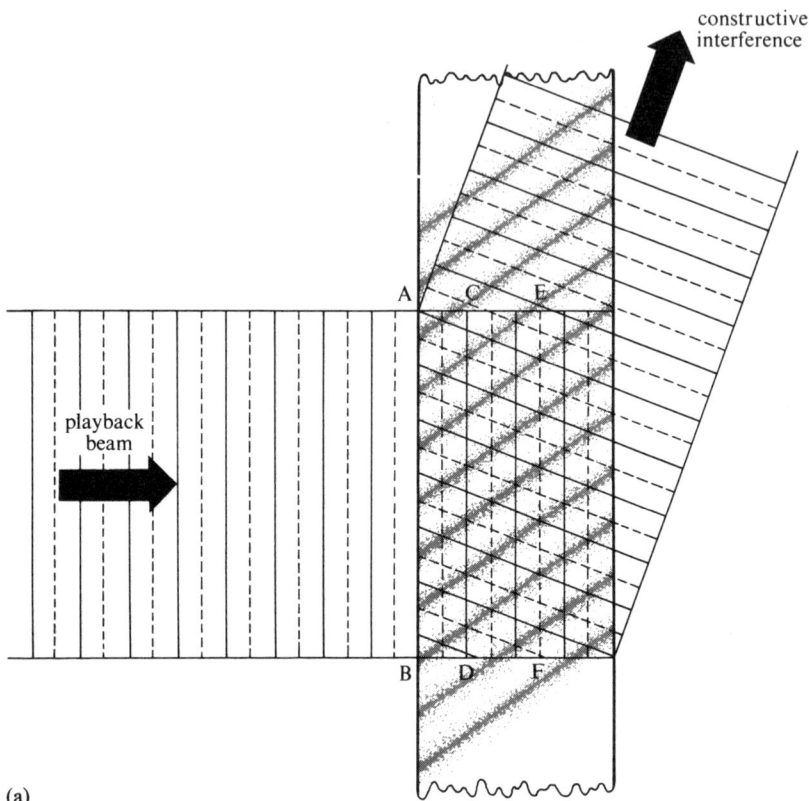

constructive
interference

constructive
interference

playback
beam

A C E

B D F

(a)

(b)

Figure 67(a)   A thick-emulsion (volume) hologram generates, an image beam when the angle of incidence of the playback beam, and its wavelength, are the same as the angle of incidence and wavelength of the reference beam used during the recording. If a different wavelength is used for playback then the angle of incidence must also be changed to compensate for this (and vice versa). Notice from the Figure, that a crest always intersects a trough at a transmission minimum (i.e. a fringe maximum) *throughout the whole depth of the emulsion.*

(b)   You might find it useful to think of this three-dimensional diffraction problem in terms of *'reflection'* from the fringe-planes. (The phenomenon is sometimes called *Bragg 'reflection' (diffraction)*.) A strong image beam only occurs when constructive inteference takes place between the rays 'reflected' from the different planes.

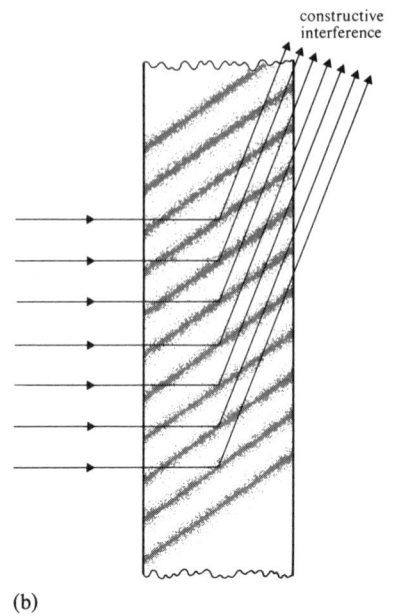

Bragg 'reflection' (diffraction)

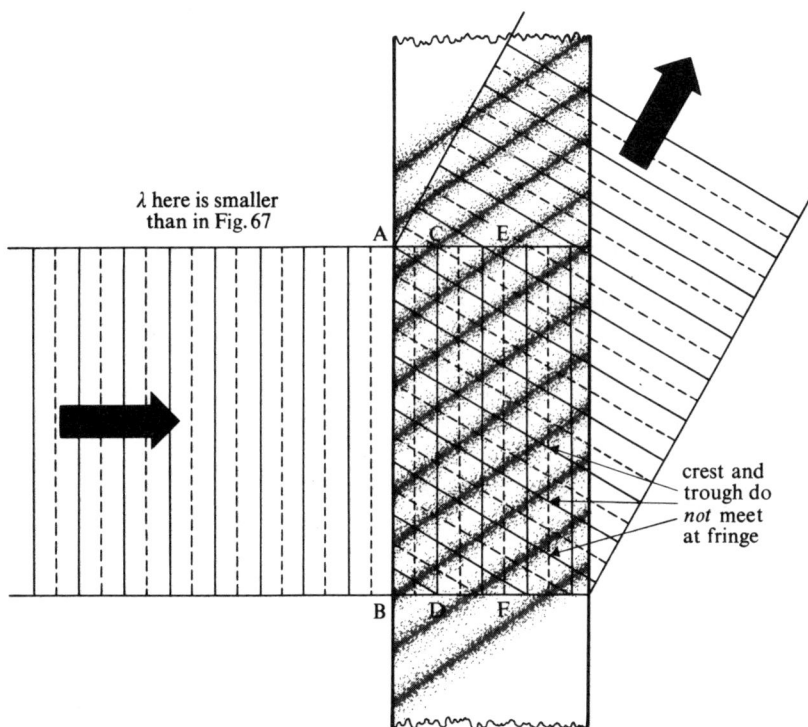

λ here is smaller
than in Fig. 67

A C E

B D F

crest and
trough do
not meet
at fringe

Figure 68   If a smaller wavelength is used to play back the volume hologram, then the output wave, since it must still satisfy the 'boundary' condition in the plane AB, will be deflected through a smaller angle than that shown in Figure 67(a). But now the wave will not satisfy the 'boundary' conditions *throughout the depth of the emulsion.* The image will be suppressed.

So, unlike two-dimensional holograms, which will diffract all colours regardless of the angle of the incident illumination, volume holograms are both angle- and wavelength-selective. If a volume hologram has been made with illumination of a given wavelength, then (assuming no shrinkage of the emulsion during development) reconstruction of the image (with the same wavelength illumination) will only occur when the playback beam is incident on the hologram at an angle identical to that of the reference beam during the recording process.*

In the same way, to reconstruct the volume hologram with different wavelength illumination, the angle of incidence of that illumination must be changed so as to again generate constructive interference along the direction of diffraction. The corollary of these facts is that if a volume hologram is illuminated with mixed-wavelength illumination (say both red and blue light) only light of that wavelength which satisfies the constructive interference condition within the depth of the emulsion will be diffracted into the holographic image. In this way the spurious images encountered with thin emulsion holograms can be suppressed. Now, only red light can replay 'red' fringes, and blue light replay 'blue' fringes. The volume hologram is behaving as its own internal colour filter.

> **SAQ 17**   Do you think that a volume hologram reconstructs both the real and virtual images equally well?

8.3   White-light volume holograms

In the volume holograms described in the previous Section, the number of fringes encountered by the light as it passes through the *depth* of the emulsion is much less than the number of fringes it encounters across the *width* of the emulsion. Consequently, the constructive interference effect will not be quite as strong as I have hitherto led you to believe—there are simply very few waves to add-up in phase. So, although using light of the wrong wavelength (or light of the right wavelength, but at the wrong angle) to playback the hologram will certainly tend to wash out the diffracted image, the wavelength (or angle) would have to be very wrong for the image to disappear altogether. If the wavelength (or angle) were only slightly different from the recording wavelength (or angle), then the intensity of the image would be only slightly reduced—the 'out-of-phaseness' of the diffracted components in this case would not be very great. In other words, the internal colour filter which the volume hologram possesses is a rather broad-band filter.

The consequence is that the kind of volume holograms described in the previous Section are not capable of regenerating a single, clear image, when illuminated with *white* light. Each of the wavelengths passed by the hologram's internal colour filter will produce an image of the object, and since different wavelengths will diffract at different angles (the diffraction-grating effect), each of these images will be slightly displaced from the other, giving rise to a 'smeared-out' rainbow-coloured multiple-image.

The way to remedy this situation is to narrow-down the number of wavelengths which can be passed by the volume hologram, i.e. to make the internal colour filter a narrow-band filter. To do this, we must make holograms which have a greater number of diffracting planes in the depth of the emulsion, so that there are more contributions to the in-phase wavefront progressing in the diffraction direction. Then, the diffracted image is washed out much more quickly as the angle of incidence of the playback beam (or its wavelength) is allowed to deviate from the angle (or wavelength) of the reference beam used in the recording.†

There are two ways of increasing the number of diffraction planes in the depth of the emulsion. We either thicken the emulsion, or we close up the planes. Since the

---

*You may even have noticed this effect to a small extent with your own 'home-made' holograms. You may have found that the brightness of the holographic image can be optimized by adjusting the angle at which the playback beam strikes the film plane.

†An analogous process occurs with two-dimensional diffraction. If a beam of light is diffracted by *two* narrow slits, the far-field diffraction pattern is basically cosinusoidal. As the number of slits in the diffraction mask is increased, the diffraction angles become much more well-defined. In the limit, a diffraction grating, with a very large number of slits, diffracts light only at well-defined angles, giving spots of light in the diffraction pattern. In summary, the more diffracting elements we have, the more directional is the diffracted light.

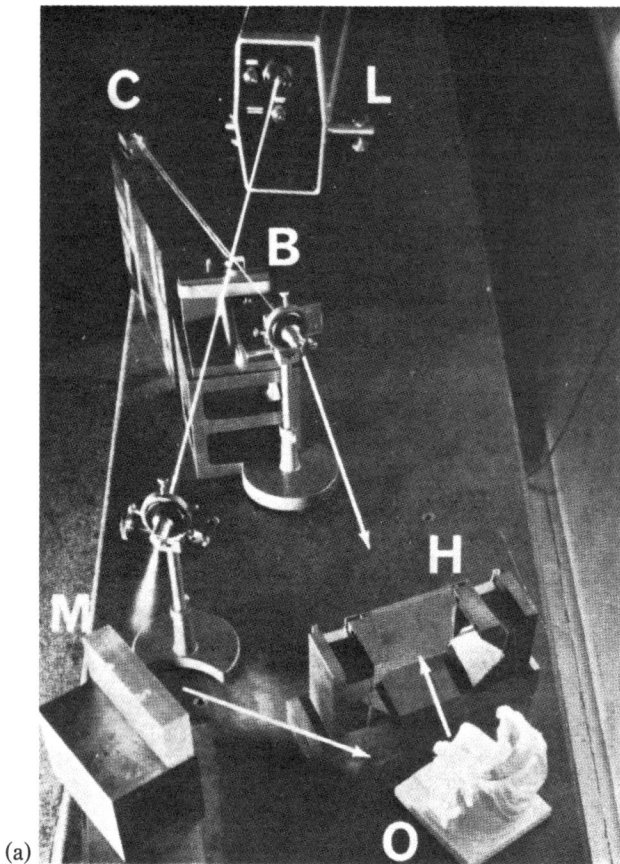

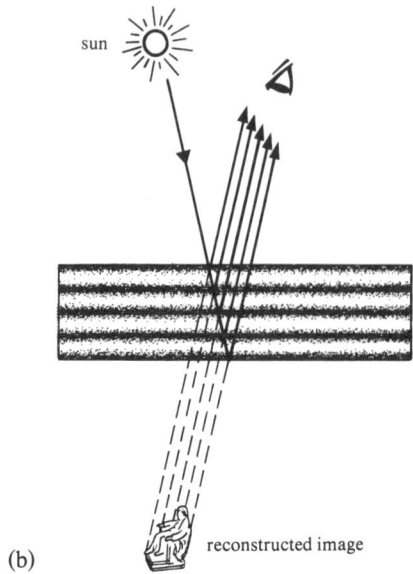

(b)

(a)

Figure 69 (a) A white-light reflection hologram is made by applying the object and reference beams from opposite sides of the photographic plate (L = laser, B = beam-splitter, C = corner-cube reflector (which reflects the light back along its own path), M = mirror, O = object, and H = holographic plate.) (b) This generates diffraction planes lying parallel to the plane of the plate, and spaced by about $\lambda/2$. The holographic image can then be reconstructed with reflected white light illumination. (c) The reconstructed holographic image.

(c)

plane-spacing is inversely proportional to sin $(\theta/2)$*—where θ is the angle between the normally-incident reference beam and the object beam—we can close up the planes by increasing the angle θ. In the limit, this angle becomes 180 degrees, and the reference and object light have to be introduced from opposite sides of the photographic plate (Fig. 69). In this situation the interference planes are approximately parallel to the surface of the emulsion, and are separated by about $\lambda/2$. The emulsion is recording the peaks of the standing wave generated by the two interfering beams (Fig. 70).

*Perhaps you'd like to try proving this for yourself. Look again at Figure 66, remembering that sin $2X = 2$ sin X cos X.

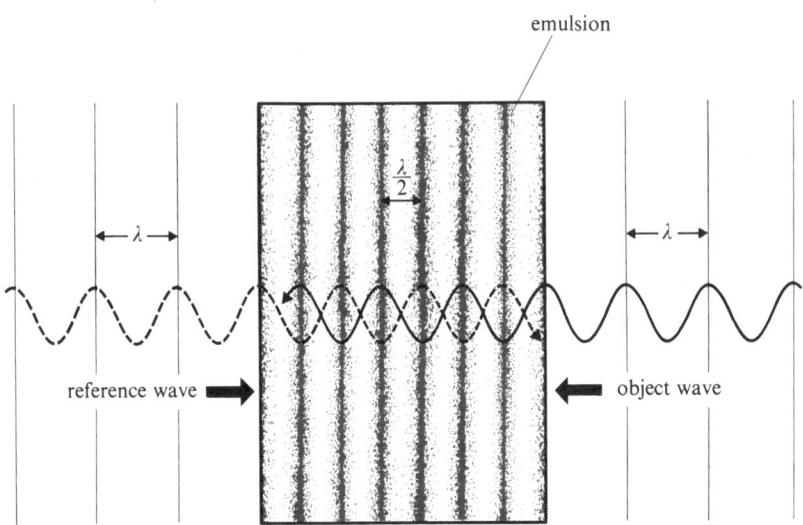

Figure 70 The diffraction planes in a white-light hologram are generated at the maxima of the standing wave.

Since holograms made in this way are viewed in *reflected* light (remember, diffraction gratings work as well in reflection as in transmission), they are frequently called *reflection holograms*. The internal colour-filter is now sufficiently narrowband for these holograms to be viewed in *white light* (provided that it is reasonably spatially coherent). The sun is a common source of such light, but a projector lamp, or an ordinary tungsten bulb, may also be suitable. Furthermore, if three beams of coherent light, at three discrete wavelengths (one for each of the primary colours) are used to generate the hologram, then white-light illumination will yield a full-colour image. That's quite an advance; full-colour, three-dimensional images, which can be viewed in sunlight—and all stored on 'black-and-white' film!

reflection hologram

SAQ 18   A reflection hologram is made using the red light from a He-Ne laser. It is then viewed in the white light of a pen-torch. Describe the appearance of the reconstructed image.

SAQ 19   In practice, it is difficult to prevent some shrinkage of the emulsion during development of the hologram. What effect would such shrinkage have on the reconstructed image formed by the reflection hologram described in the previous SAQ?

9.0   Holography: Recent advances — *Where now?*

9.1   Your guess is as good as mine

It is always a risky business prophesying what is, or is not likely to happen in the field of scientific discovery and application.* The only prediction I know to be infallible is the one which says that the more you stick your neck out—no matter whether for or against the likelihood of a particular development—the more probable it is that you will get your head chopped-off!

So I intend to content myself, in this last Section, with a quick look round at some of those research areas in holography which are currently exciting interest or speculative discussion. I can then only hope that those areas which do eventually become important, will at least have been mentioned here.

For several years now, holography has had the reputation of being a 'solution in search of a problem'. Indeed it is quite true to say that holography did not immediately live up to the expectations of the early researchers in the field. Perhaps they were a trifle over-enthusiastic in predicting that major breakthroughs, in a multitude of scientific areas, would be brought about by the application of holographic techniques. But perhaps they can be forgiven their enthusiasm; holography is certainly a technique to capture the imagination. And now, even if a bit belatedly, it does seem to be starting to live up to its early promise.

9.2   Holographic interferometry

In these two Units, I have concentrated solely on the *imaging* capabilities of holography, because that is, after all, largely what this Course is about. Yet it may well be in the field of *non*-imaging applications, that the potentially most valuable benefits of holography are to be reaped. For instance, the very important technique of *holographic interferometry*† has already been used to non-destructively test for hidden defects in car tyres, for strain contours on jet-turbine blades, to study the

holographic interferometry

---

*Lord Rutherford once said, 'The energy produced by the breaking down of the atom is a very poor kind of thing. Anyone who expects a source of power from the transformation of these atoms is talking moonshine.' These comments are totally irrelevant to Units 9 and 10, of course, but they do serve to show just how wrong you can be!

†The details of *holographic interferometry* are explained in TV programme 10, and so will not be given here. Suffice it to say that the technique involves the superposition on the same plate of two or more holograms of the object under study. Any movement of the object between the several holographic exposures will then show up as interference fringes distributed across the reconstructed image. These fringes can then be analysed to provide information about the properties of the object.

vibration patterns of musical instruments and loudspeaker cones, and the shock waves created by a bullet in flight. (See Figs. 71(a)–(d)). It has even been used in the medical field to make visible the peculiarities of human chest motion! The great advantage of this technique, in the majority of these cases, is that the same information was not previously obtainable (at least not in such detail) by other methods.

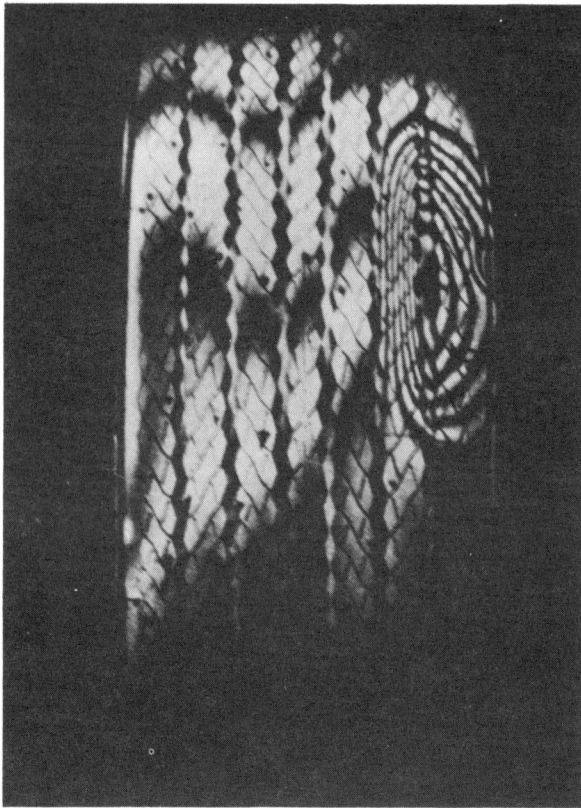

(a)

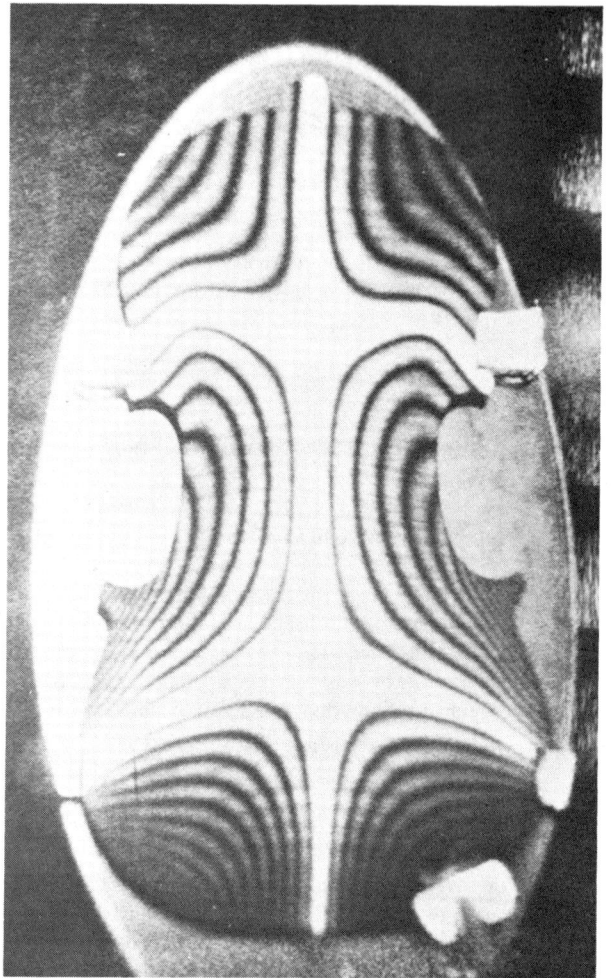

(b)

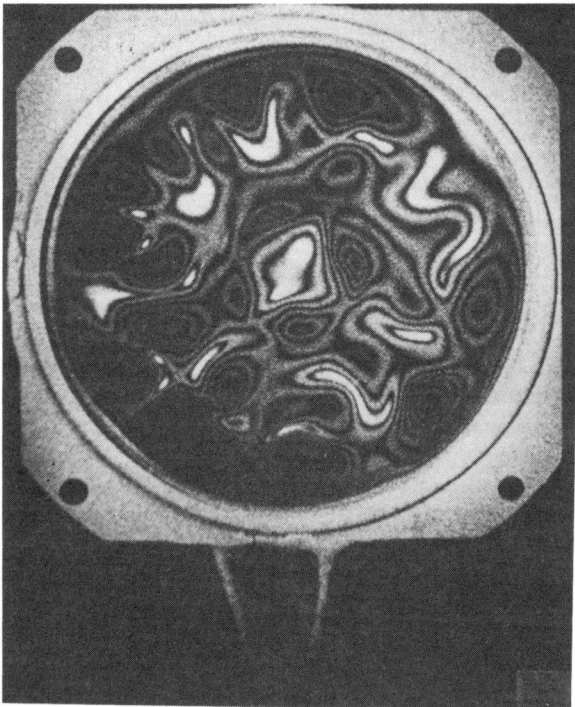

(c)

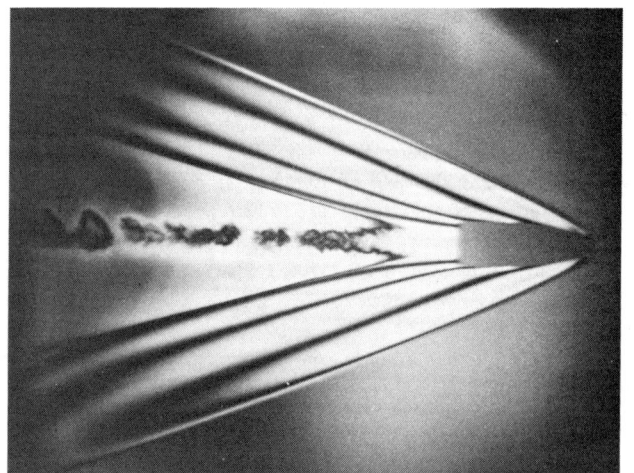

(d)

Figure 71 Holographic interferometry used— (a) to detect defects in car tyres;   (b) and (c) to observe vibration patterns on a violin and a loudspeaker cone; and   (d) to study shock waves associated with a bullet in flight.

9.3   Holographic data-storage and retrieval

Another application which is exciting much interest at the moment is the use of holograms as *data-storage* media in computers. We saw in Section 8.2 that volume holograms will only reconstruct the holographic image if the replay beam is incident on the hologram at exactly the correct angle. We can take advantage of this fact to 'multiplex' several different holograms onto the same holographic plate by using a slightly different reference angle for each hologram. Then, on replay, only that hologram corresponding to the angle of incidence of the read-out beam will be reconstructed; change the angle of this beam, and a different hologram is selected. A single crystal of lithium niobate, for example, can be made to store several thousand holograms in this way. Read-out of any one of these holograms is then achieved by addressing the crystal with a laser beam incident at the appropriate angle. Can you imagine a whole library of books, or the whole country's income-tax returns, or Scotland Yard's complete fingerprint library, stored on a handful of small transparent crystals?

holographic data storage

Actually, the institution of a holographically-stored fingerprint library is not, of itself, a very useful advance. The difficulty with fingerprints is not their storage, but the matching of the 'straight-from-the-murder-weapon' fingerprint, with the correct print in the store. But here again holography can help, with the technique known as *character-recognition*. This requires that a hologram first be made of the Fourier transform of the unidentified fingerprint. This hologram can then be used as a complex* spatial filter in an imaging system. The suspect fingerprint can then be rapidly identified simply by presenting a whole library of fingerprints to the imaging system and looking to see which of these prints produces a Fourier transform field that exactly matches the Fourier transform field stored on the complex filter. (More about this in Radio 5.)

character recognition

9.4   Computers and holograms

The love-affair between computers and holograms is not a one-sided business. If it is true to say that holography can be of help to the computer industry, it is equally true to say that computers can be of service in the making of holograms. With the computing power of a large digital computer, it becomes possible to calculate the point-by-point irradiance distribution which a fictitious object (plus reference beam) *would* have generated at the holographic plate in a hypothetical recording session. If this calculated fringe pattern is then read out from the computer onto, for example, a video-display unit, and the display subsequently photographed, it is possible to use the developed photographic negative as a real hologram, and so regenerate a three-dimensional image of an object which never actually existed!

computer holograms

9.5   Holography in commerce and entertainment

But interesting and exciting though these specialist applications may be, it is in the field of straightforward three-dimensional imaging that the greatest popular interest lies. Work is already in progress to develop three-dimensional advertising displays which do not require the highly-coherent light of a laser to regenerate the holographic image. Imagine the effect of an apparently solid hand extending from the jeweller's shop-window, tantalizingly dangling a priceless diamond bracelet in front of your very eyes! This hologram exists, and has been on display in the USA. Or then again, consider how easy it would be to provide theft-proof displays of rare or precious art objects using holographic techniques. (Perhaps some of you saw the holographic display of Thracian coins—part of the Exhibition of Thracian Treasures from Bulgaria—which was on show in the British Museum in 1976. The quality of the holographic image in that case was not as high as we might have liked, but who knows what improvements future developments might bring?)

A considerable amount of effort has also gone into producing *movement* of holographic images. Some success has already been achieved. The obvious technique is to make a sequence of holograms of an object, with a small movement of the object between successive exposures—rather as in conventional movie photography. Sequential viewing of the holograms would then give a sense of

holographic movies

---

*That is, including both phase and amplitude information.

motion. Suppose, for instance, that when you made your set of 35 mm holograms using the holobox, you had rotated the object-table slightly between successive exposures, instead of simply experimenting with different exposure times. Then, on reconstruction, as you pulled the 35 mm film-strip through the filmholder, you would have generated the illusion of movement.

There is another, rather different technique which has also been used to generate apparent motion of the three-dimensional image. It takes advantage of the fact that a thick-emulsion holographic film can store the fringe information corresponding to a number of independent, yet sequential images, simply by using a slightly different reference-beam angle for each of the fringe patterns. (Refer back to Sections 8.2 and 9.3.) Motion of the image is then effected on playback by slowly changing the angle of incidence of the playback-beam at the hologram, so revealing each scene in sequence.

The drawback with both these techniques is that the virtual image must be viewed *through* the holographic plate. This severely limits the size of the potential audience, and hence detracts from the commercial viability of holographic movies.

> **SAQ 20** Why don't we project the image in front of the film screen (by utilizing the real, rather than the virtual image) to make holographic cinema available to a much larger audience?

You might think, therefore, that *holographic television* would be the obvious candidate for commercial exploitation. And so it might, when one or two of the major stumbling blocks have been overcome.

holographic TV

Perhaps the major problem is the one associated with the massive information content of a typical hologram.

> Suppose, for example, that the hologram which is to form the TV screen is to be 30 cm square. Assuming a typical holographic fringe resolution of 1 000 lines per mm, how many information elements would the hologram contain?

It is not difficult to calculate that, in the example quoted above, 3×10^5 fringes (1 000 lines per mm $\times$ 300 mm) are required in the horizontal direction, and an equal number in the vertical direction. This means that there are about 9×10^{10} information elements in a typical 30 cm square hologram. If this hologram is to be changed 50 times per second (as with conventional TV pictures), the television system must be capable of transmitting something of the order of 45×10^{11} bits of information *per second*. From our experience of the present TV system, we know that, as a good rule of thumb, the bandwidth associated with a given TV channel must be about one-sixth of the information rate. That is, we would need a typical bandwidth of between 10^{11} and 10^{12}Hz to cope with a three-dimensional, holographic TV system. Compare this with the current TV bandwidths of a few megahertz. Holographic TV would appear to require carrier frequencies in the visible-light part of the spectrum! But with the development of lasers which can have their beam intensities modulated, and the improvements in optical fibre cables, such super-wideband optical communication links could possibly become quite commonplace in the not too distant future. In the meantime, a more prudent line to follow would be to look for ways of reducing the amount of information which has to be handled. One such way (which is currently being investigated) would be to dispense with the *vertical* parallax in the reconstructed holographic image. This would not seriously compromise the three-dimensionality of the scene (it would seem that we tend to move out heads more frequently from side-to-side, than up-and-down), but it would bring the information content just about within the reach of present communication systems technology. There is however, another major problem which has to be solved before holographic TV can become a reality. If the TV image tube is to display fringes of sufficient resolution to reconstruct either the real or virtual image, and if these fringes are to be updated at the rate of 50 frames per second, then we are going to require new materials on which to 'print' these fringes—and not just print once, but print and erase and reprint for the duration of the life of the TV receiver. Some promising new materials (photochromic and thermoplastic materials) do exist, but there is still a lot of work to be done before they become competitive with the best photographic plates presently available for conventional holography.

So, everything considered, holographic TV looks like being nothing more than the gleam in a few scientists' eyes for quite some time to come. Nevertheless, I don't want to sound too pessimistic. As I said at the beginning of this Section, crystal-ball gazing into the scientific/technological future is a very risky business—and I would like to hang on to my head for just a bit longer!

Further reading

If you would like to find out more about holography—particularly some of the more modern techniques discussed in the radio and TV programmes—then you will probably find the following books and articles helpful. In the main they are all eminently readable.

Books

Leymann, M. (1970) *Holography: Technique and practice,* Focal Press Ltd

Hecht, E. and Zajac, A. (1974) *Optics,* Section 14.3, Addison-Wesley Inc.

Articles

Leith, E. N. and Upatnieks, J. (June 1965) Photography by laser, *Scientific American,* **212,** No. 6, 24–35

Ennos, A. E. (1967) Holography and its applications, *Contemp. Physics,* **8,** 153–170

Pennington, K. S. (Feb. 1968) Advances in holography, *Scientific American,* **218,** No. 2, 40–48

Leith, E. N. (Oct. 1976) White-light holograms, *Scientific American,* **235,** No. 4, 80–95

SAQ answers and comments

SAQ 1 If we call the eyelens-retina distance v, and the eyelens-object distance u, then the height of the retinal image h is given by

$$h = \frac{v}{u} \times 5 \text{ cm} \qquad \left(\text{Magnification} = \frac{v}{u} \text{ from } WR\right)$$

Therefore
$$h_1 = \frac{5v}{25} = \frac{v}{5} \text{ cm}$$

$$h_2 = \frac{5v}{1000} = \frac{v}{200} \text{ cm}$$

Thus, the retinal image is reduced by a factor of **40**.

SAQ 2 You shouldn't have found it too difficult to find several things in the drawing which would benefit from modification. I don't intend to give you an exhaustive list of such items here (you'll probably be able to find more than I did anyway!), but I will try to identify some of the major points you should have spotted.

1 The figures are all the same size, irrespective of their distance from you, the viewer. The size of the background figures should be reduced.

2 The *shape* of the dug-out hole in the hillside is wrong. It should be distorted (flattened) much more to accord with our viewpoint at the bottom of the hill.

3 Parallel lines should converge towards a single point as they recede. There is something very wrong with the two edge-poles which form the base of the roof on the hut in the left of the picture.

4 Shadows are depicted in the scene (round the edge of the hole, at the top of the glass-furnace and on the cliff-edge, for instance) but they do not seem to be produced consistently. It is very difficult to determine the supposed source of light.

And so on—I'm sure you've got the idea by now.

SAQ 3 The same sort of comments apply here as applied in the answer to SAQ 2. A few specific points are listed below.

1 The 'parallel' escalator hand-rails appear to converge as they recede.

2 The human figures appear to be smaller at the (distant) top of the escalator, than they do at the (near) bottom of it.

3 People in the foreground 'obscure' people in the background.

4 Even with relatively diffuse lighting, shadows still give 'depth' to people's faces.

SAQ 4 1 Parallax cues (which depend on head movement) are not present on conventional photos. Using *movie-film* can help restore these cues, but this is not as good as being present yourself at the live-viewing situation. The movie-film throws away that information about head movement which your neck muscles can provide you with when you are there looking at the object itself.

2 Stereoscopic vision cues are not present on conventional photos. Stereoscopic photography can remove this deficiency more-or-less perfectly.

3 A conventional photograph is positioned in a single plane. Consequently only one eye-focus position is required to see all the items in the photograph. If the photograph does depict a three-dimensional scene, such a state of affairs is unnatural—your eyes would have to change their focus in the live-viewing situation to see clearly objects at different distances. Some allowance can be made for this when taking the photograph, by keeping the depth of field small (so that much of the photograph is out of focus). This does provide a depth cue, but unfortunately it throws away a lot of other information about the objects not in focus.

SAQ 5 The answer to both parts of this question is 'Yes'. You can always get back what you started with. That's what we mean by a perfect image, in the more normal sense of the word. (We use the word 'image' rather more generally in this Course—as you've noticed!) In the case of amplitude-only information, high-fidelity visual recording and reproduction can be achieved by simple photographic procedures. After all, a photograph of an amplitude grating could be made indistinguishable from the real amplitude grating. But then an amplitude-only object can never be three dimensional, and as such it hardly comes into our discussion of realism in visual reproduction.

SAQ 6 All phase ambiguity could be removed by adding the coherent background *in phase* with the central lobe of the sinc function. In this case a much smaller background is required, as you can see from Figure 72. The combined diffraction field shown, has an amplitude which is everywhere positive.

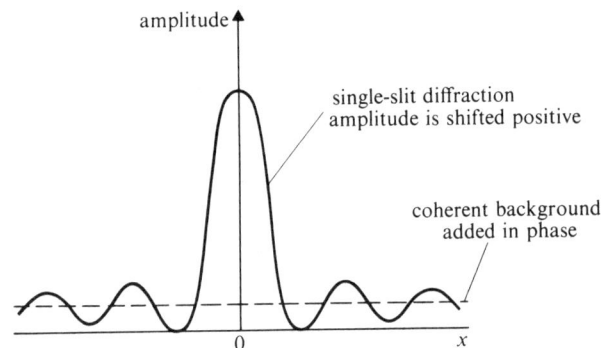

Figure 72

SAQ 7 At E both contributions to the resultant wave are in phase. As we move across the screen towards E′, the wave which followed the route striking the mirror DD′ falls progressively behind the wave which was reflected by the mirror CC′. When we reach the first dark fringe on the viewing screen (as we move left from E), we know that the optical path difference between the two waves has increased by $\lambda/2$ (the waves are in antiphase). When we reach the next bright region of the viewing screen, the waves must be in phase again, and the optical path difference must now have increased by λ.

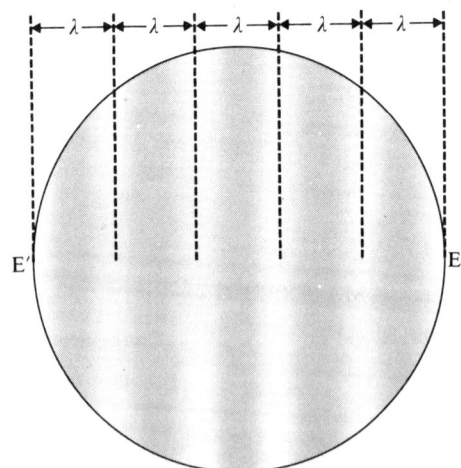

Figure 73

If you look at Figure 73, you can see that every movement across the screen from a bright region to a bright region, corresponds to an extra difference of λ in the optical paths of the two beams. From the Figure, it is clear that the optical path difference between the two beams at E′ is greater than the optical path difference between the two beams at E, by a distance 5λ. But the *optical* path difference is twice the *physical* path difference (the light returns along its own path). So D′ must be a distance **2.5 λ** (≈ 1.6 μm if He-Ne light is used) behind D.

SAQ 8   The spatial frequency of the fringes on the grating is given by

$$q = \frac{\sin \theta_1 - \sin \theta_2}{\lambda} \qquad \text{(from equation 10)}$$

But $\theta_1 = \theta/2$

and $\theta_2 = -\theta/2$

(the angles must be measured in the same sense).

Hence when the two beams strike the photographic plate symmetrically from each side of the normal, the expression for the spatial frequency of the fringes becomes

$$q = \frac{2 \sin \theta/2}{\lambda}$$

In the case cited in the question, $\theta = 50°$ and $\lambda = 633$ nm

Therefore $q = \dfrac{2 \sin 25°}{633} \times 10^9 \text{ m}^{-1}$

$$= \frac{0.845 \times 10^9}{633} \text{ m}^{-1} \approx 1\,335 \text{ cycles mm}^{-1}$$

SAQ 9   The answer to this question is, 'Yes, it would.' But this does *not* spoil the theory, as you might at first think. If you consider viewing a three-dimensional transmission object, say a glass tumbler for instance, then it is quite true that the optical field carrying information about the back of the tumbler is distorted as it passes through the rest of the tumbler. (You'd be very surprised if you could look through the 'front' of a glass object, and see the 'back' of the object as undistorted as it would be if the front part were removed altogether). This information must be carried over to the hologram plane, where it appears in the form of more complicated fringes. Similarly, if any part of the plane P is obscured by one of the planes Q or R from some view point on the hologram, then that light will not reach the hologram at that point. The fringes corresponding to the light from that plane will be absent from the hologram at that point. Hence the hologram recording will exactly mimic the real-live viewing situation, and the reconstructed image will appear exactly as the real-live object did.

SAQ 10

(i)   If $I_R = 2I_{OB}$

then $A_R^2 = 2A_{OB}^2$

and $A_R = \sqrt{2}\, A_{OB}$

So $I = \dfrac{2A_{OB}^2}{2} + \dfrac{A_{OB}^2}{2} \pm \sqrt{2}.\, A_{OB}^2$

$= 0.1\, A_{OB}^2$ to $2.9\, A_{OB}^2$

Therefore the *depth of modulation* $= \left(\dfrac{2.8}{3.0} \times 100\right)\% = 93.3\%.$

(ii)   If $I_R = 9I_{OB}$

then $A_R^2 = 9A_{OB}^2$

and $A_R = 3A_{OB}$

So $I = \dfrac{9A_{OB}^2}{2} + \dfrac{A_{OB}^2}{2} \pm 3A_{OB}^2$

$= 2A_{OB}^2$ to $8A_{OB}^2$

Therefore the *depth of modulation* $= \left(\dfrac{6}{10} \times 100\right)\% = 60\%.$

SAQ 11   The lettering is mirror reversed because you are looking at the rays that are 'reflected' back toward you by the white card—that is, you are now looking away from the laser, rather than towards it. To see the lettering the correct way round, you should project the real image onto the ground-glass screen, and look at this projected image from the side of the screen *away* from the laser, i.e. look *towards* the laser. The projected images from your home-made holograms will probably be too faint for you to see them in this way; this approach, however, works very well with the professional hologram. You are asked to view the professional hologram in this way in Section 7.

SAQ 12   To check that your hologram really is a sinusoidal grating, hold it in the path of the undiverged laser beam. The hologram should then diffract the laser beam in a way characteristic of a sinusoidal grating. That is, the single input beam to the hologram should be split into *three* output beams—one going in the same direction as the input beam, and the other two being diffracted each side of this beam by an angle θ. This angle should be equal to the angle between the reference beam and the 'object' beam (at the holographic film) in the recording situation.

SAQ 13   A large diameter laser beam incident on the hologram generates a real image with a *small* depth of focus; a small diameter laser beam generates a real image with a *large* depth of focus. You should notice that this relationship is identical to that which exists between the depth of focus on a photograph, and the size of aperture stop in front of the camera lens. This is not too surprising, since the reasoning in both cases follows along very similar lines.

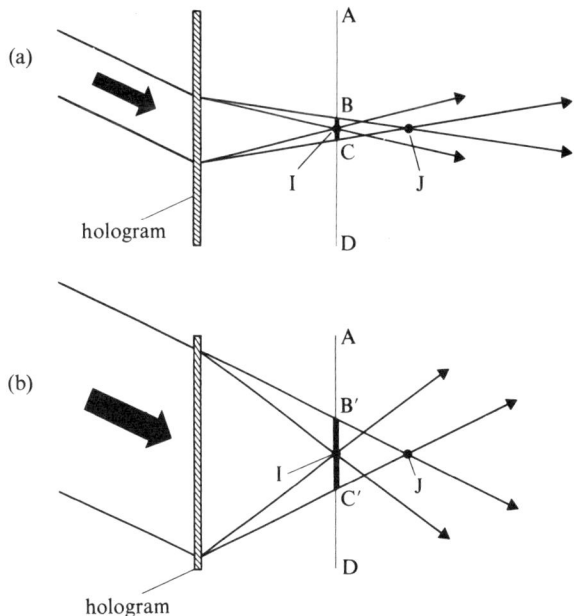

Figure 74

Look at the two diagrams in Figure 74. In (a), the hologram is played back with a narrow beam, in (b) with a wide beam. In both cases, I have shown the focus positions of two image points, I and J, corresponding to points at different *depths* in the object. (I, for instance, might be a point on the magnifying glass, and J a point on the stopwatch). If I place a viewing screen in the plane AD, then in the narrow-beam situation, I will be perfectly in focus, whereas J will be smeared out over the smallish area BC. As you can see from (b) though, when a wide beam is used to play back the hologram, point I is perfectly in focus in plane AD, but point J is smeared out over the much bigger range B′C′. That is, J is much more out of focus in (b) than it is in (a). The wider the playback beam, the smaller will be the depth of focus.

SAQ 14   The explanation is quite simple. The magnifying glass appears to be *behind* the stop-watch (even though the stopwatch is still magnified by the glass) because the real image is *pseudoscopic*.

SAQ 15 Remember that a hologram is rather like a diffraction grating. Since $d \sin \theta = \lambda$ (for any given set of fringes characterized by d) then clearly $\sin \theta$ is proportional to λ. That is, long wavelength light will be deflected through a larger angle than will short wavelength light. If holographic fringes are generated by light of a certain wavelength, and played back with longer wavelength light, then the image will be displaced because the larger wavelength light will be deflected through too large an angle. Conversely, playback with light of a wavelength shorter than that of the recording light will cause the image to be displaced in the opposite direction. The playback beam will be deflected through too small an angle.

Let me deduce this result mathematically. In Figure 65, the interference between, say, the red reference and object beams gives rise to 'carrier' fringes with a spacing which is characterized by the average angle between the two beams, and the wavelength of the red illumination. That is

$$d = \frac{\lambda_r}{[\sin \theta_r]_{record}}$$

If these fringes are now illuminated with blue light, the reconstructed angle must also change to satisfy the equation

$$d = \frac{\lambda_b}{[\sin \theta_b]_{playback}}$$

That is,

$$[\sin \theta_b]_{playback} = \frac{\lambda_b}{\lambda_r} [\sin \theta_r]_{record}$$

Hence, the average deviation for the spurious blue image (blue light replaying 'red' fringes), is *less* than for the full-colour image. Similarly, if red light plays back 'blue' fringes, then

$$[\sin \theta_r]_{playback} = \frac{\lambda_r}{\lambda_b} [\sin \theta_b]_{record}$$

so that the deviation for the spurious red image is *greater* than for the full-colour image. This was shown in Figure 65.

SAQ 16 There will be *six* spurious images. Suppose the three colours are labelled A, B and C. Then the full-colour superposed image will result from colour A replaying fringes A, colour B replaying fringes B, and colour C replaying fringes C. We can denote this as AA + BB + CC, where the first letter represents the colour of the replay beam, and the second letter represents the 'colour' of the recorded fringes. Hence, the other images will result from the combinations: AB; AC; BA; BC; CA; and CB; a total of six spurious images.

For the mathematically inclined, a hologram recorded and played-back with n colours would regenerate *one* image with the correct colour mix, and $n(n-1)$ spurious images.

SAQ 17 In the situation shown in Figure 67, the real image is suppressed; the downward travelling wave does not intersect the fringes in the depth of the emulsion at the right points to generate constructive interference. However, the real image *can* be generated by playing in the playback beam *from the other side of the plate*. In Figure 67, the reconstructing beam would be travelling from right to left, and the real image would be generated at the bottom left of the figure. In this case, the virtual image is suppressed. You might find this idea easier to grasp in terms of the 'reflection-planes' analogy shown in Figure 67(b).

SAQ 18 Assuming no shrinkage of the emulsion during development of the hologram, the reconstructed image will appear 'He-Ne red', even though viewed in white light. The other wavelengths will be suppressed. A full-colour image cannot be generated because full-colour information was never recorded. Only 'red' fringes were recorded.

SAQ 19 If shrinkage occurs during development, the quasi-parallel interference planes in the reflection hologram will get closer together. Since the planes are separated by approximately $d = \lambda/2$, a reduction of d means that reconstruction of the image will occur only with light of a shorter wavelength. That is, the colour shifts towards the blue end of the spectrum. If the hologram was recorded in He-Ne red light only, and emulsion shrinkage occurs, then *no* image will be reconstructed if only He-Ne red light is used for playback. If 'white' light is used to replay the hologram, however, the reconstructed image will appear yellow or green. Nowadays, shrinkage of the emulsion can be prevented by suitable treatment of the photographic plate after development.

SAQ 20 The real image is not really visible to any more people than is the virtual image. To see the real image, the viewer must be *within* the diverging light cone proceeding from that image (see Fig. 75). It is not possible to see this image from a position outside this cone of light. To increase the angle of divergence of the light, and hence make the image accessible to a wider audience, we must increase the size of the hologram. Very large holograms are technologically difficult to produce. And anyway, increasing the size of the hologram would also make the virtual image accessible to a larger audience!

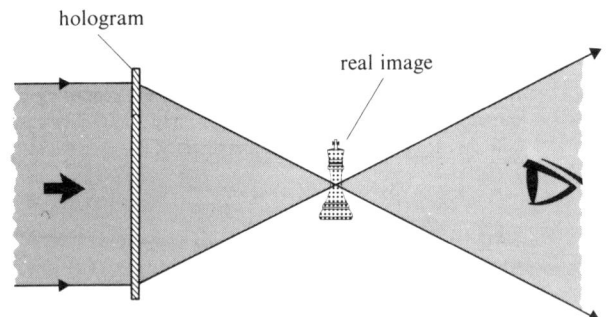

Figure 75

Exercise answers and comments

Exercise 1 Substituting the expressions for ψ_R and ψ_{OB} into equation 17 gives

$$I = \langle [A_R \cos(\omega t + \varphi_R) + A_{OB} \cos(\omega t + \varphi_{OB})]^2 \rangle$$

$$= \langle A_R^2 \cos^2(\omega t + \varphi_R) + A_{OB}^2 \cos^2(\omega t + \varphi_{OB})$$

$$+ 2A_R A_{OB} \cos(\omega t + \varphi_R) \cos(\omega t + \varphi_{OB}) \rangle$$

Using the identity

$$\cos(X + Y) + \cos(X - Y) = 2 \cos X \cos Y$$

to expand the third term in this expression, we have

$$I = \langle A_R^2 \cos^2(\omega t + \varphi_R) + A_{OB}^2 \cos^2(\omega t + \varphi_{OB})$$

$$+ A_R A_{OB} \cos(2\omega t + \varphi_R + \varphi_{OB}) + A_R A_{OB} \cos(\varphi_R - \varphi_{OB}) \rangle$$

But as you can see by referring back to Figure 18, the time-average of both the cosine squared terms is $\frac{1}{2}$, and the time-average of the $\cos(2\omega t + \ldots)$ term is zero. The $\cos(\varphi_R - \varphi_{OB})$ term does not include a time-dependent factor.

Hence $\quad I = \dfrac{A_R^2}{2} + \dfrac{A_{OB}^2}{2} + A_R A_{OB} \cos(\varphi_R - \varphi_{OB})$ $\qquad$ (Eq. 18)

This is the expression quoted in the text.

Exercise 2 If the playback wave has the same phase as the reference wave, we can write

$$\psi_P = A_P \cos(\omega t + \varphi_R)$$

Furthermore, if the amplitude transmission of the plate is proportional to I, then the transmitted wave (ψ_T) is proportional to $\psi_P I$.

Hence

$$\psi_T = A_P \cos(\omega t + \varphi_R) \left[\frac{A_R^2}{2} + \frac{A_{OB}^2}{2} + A_R A_{OB} \cos(\varphi_R - \varphi_{OB}) \right]$$

$$= \frac{(A_R^2 + A_{OB}^2)}{2} \psi_P + A_P A_R A_{OB} [\cos(\omega t + \varphi_R) \cos(\varphi_R - \varphi_{OB})]$$

Using the identity $\cos(X + Y) + \cos(X - Y) = 2\cos X \cos Y$, the expression for ψ_T becomes

$$\psi_T = \frac{(A_R^2 + A_{OB}^2)}{2} \psi_P + \frac{A_P A_R A_{OB}}{2} \cos(\omega t + \varphi_{OB})$$

$$+ \frac{A_P A_R A_{OB}}{2} \cos(\omega t + 2\varphi_R - \varphi_{OB}) \qquad \text{(Eq. 19)}$$

which is the expression quoted in the text.

Acknowledgements

Grateful acknowledgement is made to the following sources for illustrations used in these Units.

Figure 2 The British Library; *Figure 4* Courtesy of Museum Boymans—van Benningen, Rotterdam; *Figure 5* Pinacoteca di Brera, Milan. Photo by SCALA; *Figure 6* from J. L. Locher, *The World of M. C. Escher,* Harry N. Abrams Inc. N.Y., by permission of the Escher Foundation, Collection Haags Gemeentesmuseum—The Hague; *Figure 7* Louvre Museum; *Figure 10* London Transport, Photo by Heinz Zinram; *Figure 15* David Shaw; *Figure 25* From F. Zernike, *Proc. Phys. Soc.,* **61** Pt. 2, 1948; *Figures 43 and 63* from E. N. Leith and J. Upatnieks, Photography by Laser, *Scientific American,* June 1965; *Figure 50* Courtesy of L. D. Siebert, K.M.S. Fusion Inc.; *Figure 61* Copyright Griffin and George Ltd; *Figure 69* from D. Gabor *et al.*, Holography, *Science,* **173**, July 2 1971. Copyright 1971 American Association for the Advancement of Science; *Figure 71 (a) and (c)* Dr. Ralph Grant, G.C. Optronics; *Figure 71 (b)* A. E. Ennos, N.P.L.—Crown Copyright reserved; *Figure 71 (d)* R. E. Brooks, TRW Systems Group.

ST 291   IMAGES AND INFORMATION